Our Forest Legacy

Other books by Chris Maser

The Perpetual Consequences of Fear and Violence: Rethinking the Future
(Maisonneuve Press, 2004)

The Redesigned Forest

Fleas of the Pacific Northwest (with Robert E. Lewis and Joanne Lewis)

Forest Primeval: The Natural History of an Ancient Forest

Global Imperative: Harmonizing Culture and Nature

Sustainable Forestry: Philosophy, Science, and Economics

*From the Forest to the Sea: The Ecology of Wood in Streams, Rivers, Estuaries,
and Oceans* (with James R. Sedell)

*Resolving Environmental Conflict: Toward Sustainable Community
Development*

Sustainable Community Development: Principles and Concepts

Setting the Stage for Sustainability: A Citizen's Handbook (with Russ Beaton
and Kevin Smith)

Vision and Leadership in Sustainable Development

*Ecological Diversity in Sustainable Development: The Vital and Forgotten
Dimension*

Reuniting Ecology and Economy in Sustainable Development (with Russ
Beaton)

Land-Use Planning in Sustainable Development (with Jane Silberstein)

Forest Certification in Sustainable Development (with Walter Smith)

Mammals of the Pacific Northwest

The World is in My Garden: A Journey of Consciousness (with Zane Maser)

Evaluating Sustainable Development: Giving People a Voice in Their Destiny
(with Okechukwu Ukaga)

Our Forest Legacy

Today's Decisions, Tomorrow's Consequences

Chris Maser

Maisonneuve Press
Washington, D.C.
2005

Chris Maser
Our Forest Legacy: Today's Decisions, Tomorrow's Consequences

© Copyright 2005
Maisonneuve Press
P. O. Box 2980
Washington, D.C. 20013

http://www.maisonneuvepress.com

Maisonneuve Press is a division of the Institute for Advanced Cultural Studies, a non-profit collective of scholars concerned with the critical study of culture and the environment.

Library of Congress Cataloging-in-Publication Data

Maser, Chris.
  Our forest legacy : today's decisions, tomorrow's consequences / Chris Maser.
      p. cm.
  Includes bibliographical references.
  ISBN 0-944624-46-4
  1. Sustainable forestry—United States. 2. Forest management—United States. I. Title.
  SD143.M36 2005
  634.9'2'0973—dc22
                                        2005000335

ISBN 0-944624-46-4 (alk. paper)

Foreword

The opening quote by Scott Nearing about the peculiar and deeply satisfying journey of the off-beat thinker told me this would be a passionate read. So it is. Chris Maser's *Our Forest Legacy* throbs with the heartbeat of one who has not lost that child-like fascination with the natural world. And this naiveté, coupled with wisdom, simply asks us to change—in hope for the future, born of anguish about the past.

This book is compelling, as well as deeply disturbing.

It made me think, hard. Chris confronted me with things I had never even been aware of, much less grappled with. For example, I fell into the trap of thinking that "forest industry" meant logging and wood processing, while ignoring all the other forest-dependent industries, such as recreation, outfitting, fishing, farming, municipal water, and so on. They clearly have as much at stake as logging and wood processing, or perhaps even more, in the determination of whether forest policies equate to sustainable forestry. This said, his book made me more aware of how deeply we are a part of forests, and how much we are diminished by our loss of native forests due to our greed. We need forests. And Chris speaks of how we can sustain them that they might sustain us.

Yet I am disturbed. Although I'm an optimist, Chris brings me very close to utter despair. I wonder whether we have the capacity to change as dramatically as will be necessary? Have forests been so fundamentally altered, in spite of their resilience, as to be beyond restoration? I am buoyed by the knowledge that most great change has been fostered by a dedicated individual or small group. Might this book be another *Silent Spring*?

I see things through the lens of a life spent in forestry; 34 years with the U.S. Forest Service. I felt a pull on my life in my early teens. Chris had water-filled ditches behind his house; I had summers with my Dad in the Black Hills of South Dakota and the Bighorn Mountains of Wyoming. Dad looked at fossils. I explored and was awestruck. So, I understand, Chris, why forestry is more than a job. Being a zealot is not a bad thing when you sense the world slipping away. Make no mistake, Chris Maser bears a torch. And we need torchbearers, for the human heart grows cold slowly and needs a fire to rekindle passion.

Gifford Pinchot dreamed of an organization—the U.S. Forest Service—being moved enough by the idealism of conservation that it would provide exemplary stewardship of the lands entrusted to its care, especially for future generations, of which we are now one. For decades the Forest Service seemed up to the lofty challenge. But today it is a leaky vessel, and voices are being heard that it is beyond saving.

The century-old notion of national forests—and a government and a people being wise enough to set aside lands dedicated to principles of conservation rather than exploitation—was radical. But this idea has stood the test of time and is broadly emulated internationally.

Nevertheless, many people have given up on the Forest Service, and I sense that Chris was also close to giving up. I'm glad, however, that he chose instead to invest hope in the concept of the Forest Service and our national, public forests, to illustrate what forestlands can mean for our future when properly understood and appreciated. His vision is a call to action.

I first met Chris in the mid-1980s while I was working on the San Juan National Forest in Colorado. He had been invited to speak to local Forest Service officials. Who was this guy talking authoritatively about truffles, mouse poop, mycorrhizae, northern spotted owls, and bugs? And making sense! But why was I hearing such "truths" for the first time, after 20 years in the profession?

I eventually became Supervisor of the Siuslaw National Forest, which is headquartered in Corvallis, Oregon, where Chris lives. I gave a speech shortly after the Northwest Forest Plan was released in 1994. The Northwest Forest Plan—"The Answer" to decades of forest conflict in the Pacific Northwest—laid out a challenge to change the path of management on federal forests (and by implication on *all* forests). Making the plan work, however, was an altogether different challenge.

When Chris read my speech a few days later, he called me and said that if I needed *anything* from him to help me implement the forest plan, to just ask. I never redeemed the favor—until now, and I'm late in asking. If I had been thinking beyond myself, I would have asked for a BIG favor, something that would benefit many people, not just me.

I would have asked Chris to write a book about why forests are important to the world; why they ought to be recognized as a biological living trust we hold for future generations; and why the U.S. Forest Service is in a unique position to lead a radical reformation. Chris, I would have asked you to put in everything you've learned about forests, and most especially what you've learned about people. But you anticipated this need—"Our Forest Legacy" is just the favor I should have asked for. I'm glad you didn't wait. We need your wisdom. May it move us to action.

Jim Furnish
Deputy Chief
National Forest System
USDA Forest Service (retired)

Dedication and Acknowledgments

There are times in the chronicles of history that people have stepped upon the stage of life and acted in ways that left a legacy through the ages. This book is humbly dedicated to the memory of two such men—Theodore Roosevelt and Gifford Pinchot—who, with foresight and courage, helped to establish our National Forests in trust for all generations and the U.S. Forest Service to oversee them.

It is with humility that I also dedicate this book to my son, Erik, who represents the future, the beneficiaries of our forest legacy. May our forest legacy serve him—and all generations—well.

It is with great pleasure that I thank for following people for making the time to thoughtfully review the manuscript and help me to struggle toward that ever-elusive horizon of excellence: Helmut Blaschke (Research Forest Biologist, Technishe Universität München, School of Forestry and Natural Resource Management, Freising, Germany), Mary Chapman (Executive Director, Forest Stewards Guild, Santa Fe, NM), Norm Christensen (Nicholas School of the Environment and Earth Sciences, Duke University, Durham, NC), David Conklin (Computer Programmer, Department of Botany, Oregon State University, Corvallis), Jim Furnish (Deputy Chief, National Forest System, USDA Forest Service, retired, Washington, D.C.), Toddi A. Steelman (Assistant Professor, Environmental and Natural Resource Policy, Department of Forestry, North Carolina State University, Raleigh), Clinton Trammel (Forest Manager, Pioneer Forest, Salem, MO), Bill Wilkinson (Senior Forester, Forest Stewardship Council-US, Arcata, CA; formerly, Forester, Hoopa Tribal Forestry, Hoopa, CA), and Michele Zukerberg (Forester, Washington Department of Natural Resources, Olympia, WA).

I owe special debt of gratitude to my lovely wife, Zane, who not only came up with the idea for this book and insisted that I write it but also was patient with me during the many hours that I worked on it. In addition, Zane spent many hours proofreading the galley.

The majority will always be for caution, hesitation, and the status quo—always against creation and innovation. The innovator—he [or she] who leaves the beaten track—must therefore always be a minoritarian—always be an object of opposition, scorn, hatred. It is part of the price he [or she] must pay for the ecstasy that accompanies creative thinking and acting.

— Scott Nearing, Economist

Preface

*No idea is so antiquated that it was not once modern. No idea is so modern
that it will not someday be antiquated.*
— Ellen Glasgow, American author

If we, as a society, learned anything from the social upheaval of the "1960s,"
it must be that we cannot unilaterally destroy either our environment or our established social order without having a viable alternative to replace what we dis-
card. I say this because, whatever we humans do in the course of our living, we
affect our environment, as well as the potential orderliness of our society. Beyond
that, both the quality of our lives as individuals and our survival as a species
depend first on the health of our shared environment and second on our collective psychological maturity as a society.

We, in the U.S. today, live in what is perhaps the most wasteful, throwaway,
consumerist culture fueled by the most avaricious economic system in the world.
In the process of acquiring ever-more material goods in our search for that elusive sense of "material security," we have become almost totally disconnected
from our heritage – a way of life that beat to the pulse of the great forests, prairies,
and rivers of this land. Nevertheless, a new paradigm of social-environmental
sustainability must be available to take the place of our old, dysfunctional, consumerist one before it can be cast out. The book in hand is a small, imperfect step
in the direction of a new paradigm I propose for the practice of forestry that shapes
the caretaking of our public forests.

Each new paradigm is built on a shift of insight, a quantum leap of intuition,
with only a modicum of hard, scientific data. Those who cling to the old way
often demand irrefutable, scientific proof that change is needed, but such proof is
never available because science can only disprove something—never prove anything. The irony is that the old way also began as the new and was also challenged to prove its worth and/or its desirability.

Although time and human effort may have proven the old paradigm of forestry to have been more "right" than its predecessor, it was still only *partially*
right. So it is with the new; it too is more "right" than the old. Nevertheless, the

new will eventually also be proven only *partially* right and in need of change.

The personal and professional trap is that any comfortable paradigm becomes self-limiting because new data cannot easily fit into the old way of thinking. A belief system grown rigid with tradition and hardened with age requires a periodic "cracking open" if a new thought-form is to enter and grow, one that moves the individual and/or the profession forward with a renewed sense of authenticity. The challenge is how to re-conceptualize our experience of life by integrating new data with old knowledge and subsequently making effective, action-oriented decisions *now*. A lack of knowledge is *not* the issue, despite the fact that it both sounds good and is often put forth as "the problem." In addition, our institutions (from colleges and universities to city, county, state, and federal agencies) steadfastly resist change.

Moving forward may be difficult for those whose belief system and personal identity are totally invested in the old paradigm because, in their perception at least, there seems no reason to change. For those who subscribe to a new paradigm, moving forward is easier, since there is something toward which to move—a new view that hints at what the profession of forestry can and must become, a view more in tune with today's ecological understanding. Yet those who harbor new ideas are no better or more "right" than those who cling to the old ways; the two views are only different because "right" is a matter of perception. And one's perception is one's "truth," although we have no way of knowing how they arrived at it.

Here, I would have you consider two things. First, if you were to ask me who or what "Chris Maser" is, I could not answer. Nor could I tell you who or what you are or what a butterfly is or a tree. Oh, I could repeat your name or give you the scientific names by which I, you, a particular butterfly, and a particular tree are recognized, but what does that really tell you? Virtually nothing because of the incredible diversity that lies in life itself. Just imagine, since that first living cell, nothing has ever again been alone on Earth. Since that first living cell, the diversity of life has literally filled the planet. And the experiment continues.

How exactly that part of creation called life began is a question as old as the first human being to ponder it. Nevertheless, the first animated cell opened not only the possibility of life and living diversity but also a whole dimension of diversity beyond our present comprehension—infinite diversity, created out of nonliving substances and living tissue as well as a combination of the living with the nonliving. Think, for example, of the vast array of marine snails, each of which makes its own peculiar shell out of nonliving materials. Without the variety of living snails, the variety of shells could not exist.

The wonder of biological diversity is the wonder of its having begun with a single living cell, or maybe even a handful of cells scattered throughout the ancient seas of the world. From that cell, or perhaps those cells, arose the longest living experiment on Earth—the genetic experiment of life. You could argue that combinations of genetic materials are really no different from the original combinations of chemicals that gave rise to chemical compounds. If you omit the spark of life from this equation, you would be correct. But that indefinable spark of life is there, and that changes everything.

Today, therefore, as I meet each living thing that shares the world with me, I see the pinnacle — the culmination — of billions upon billions upon billions of genetic experiments, all of which have taken place over millions of years, all embodied in each butterfly, each rose, each tree, each bird, and each human being, including me when I peer into a mirror. Every living thing on Earth is the apex of creation, because every living thing is the result of an unbroken chain of genetic experiments (each individual that ever lived being part of a single experiment, or even a few experiments) that began with the original living cell(s) that filled the lifeless sea with life. How can a word, such as a name or a title define, describe, or characterize such wonder?

Second, we always make the best decision we can at a particular time, under a particular circumstance, with the data on hand. This does not mean, given similar circumstances, that we would make the same decision today or in the future. It only means that it was the best decision we could make at that time.

It does not mean that others will necessarily agree with our decisions or we with theirs. It only calls attention to the fact that I must accept your decision as your best because I cannot judge. I do not know why you did what you did; I only know what you did and how that *appeared* to me.

A gentleman in the U.S. Forest Service taught me much about judgment and how people hear what we say. I was giving a speech in Spokane, Washington, about fire in forested landscapes and explaining new data and new points of view. When I was finished, he came up to me and, with a quivering chin and misty eyes, said, "I've been with the Forest Service 29 1/2 years and I'm going to retire in six months. Do you mean to tell me I've been *wrong* my whole career?"

"No sir," I replied, "I'm not telling you that at all. You did the best you could with the data you had on hand. Now, however, with much new data, we can make some different choices, different decisions than you could during your career."

Looking at him, listening to his faltering question, I realized with searing insight how incorrect we are when we presume to judge and that we are doubly incorrect when we presume to judge from hindsight. Everyone does their best within their level of understanding, which is the underpinning of their level of consciousness of cause and effect. If, therefore, we feel the need to voice another point of view, such as the need for a new paradigm, let's be gentle with one another and treat one another with compassion and dignity because we are each doing the level best we can at any given moment.

Compassion and dignity have two components in the public arena that most of us seem unaware of. They are best explained in two, short stories. While still in active research as an employee of the Bureau of Land Management, I gave a speech to sixty loggers about the necessity of changing logging practices to maintain and/or enhance the forest's ecological productivity.

When I was finished an elderly man with snow-white hair, a tan, leathery face, and a twinkle in his sky-blue eyes came up to me and said, "You may have something there, Sonny, but I jist can't figure out what it is. Mostly I think yur full of shit!"

"Why do you think that?" I asked.

"Cause," he replied, "the way I think of a forest you don't make no sense nohow."

Although I felt like I'd just been kicked in the stomach by a mule, he was entirely correct. With his philosophy, his old way of thinking, new practices were nonsense. And so I learned that new scientific knowledge is useless when it collides with an old philosophical underpinning. I've therefore done my best to marry science and philosophy in the text of this book in a way that will give birth to a more sustainable practice of forestry than now exists.

The second component of compassion and dignity is the ability to change one's thinking in a safe venue. This, too, was brought home to me by a logger, a big, burley guy who got up in the middle of a speech I was giving to a group of loggers and declared in a loud voice: "I don't have to listen to this bullshit," as he stormed out of the room.

Although I did my best to ignore the incident, it bothered me throughout the rest of my presentation. Done with my part of the program, I walked out, only to find the man waiting for me just outside the door. At first, I thought he might hit me because he looked agitated, but instead, he took me aside and said: "That was one of the best speeches I ever heard. And I agree with most of what you said, but I never thought about it much. But I couldn't let the others know 'cause they're more sawdust loggers than I am, and I still have to work with 'em."

To me, these stories point out that, whether we realize it or not, we need one another. Consider, for example, large, old trees on the side of a mountain. Each signifies primeval majesty carried forward through the centuries, but only together do they represent an ancient forest. Yet, we do see the forest when we look at the individual trees. If we could see belowground, however, we would find gossamer threads of mycorrhizal-forming fungi stretching for billions of miles through the soil. These fungi grow as symbionts on and in the feeder roots of the ancient trees. Not only do they acquire food in the form of plant sugars through the roots of the ancient trees but also they provide nutrients, vitamins, and water from the soil to the trees and produce growth regulators that benefit the trees. These symbiotic fungus-root structures (mycorrhizae) are the termini of the threads that form a complex fungal net under the entire ancient forest and connect all trees one to another.

Like the ancient trees, we are separate individuals, and like the ancient forest united by its below ground fungi, we are united by our humanity—our need for love, trust, respect, and unconditional acceptance by one another of who we are as individuals. So, in determining whose judgment in a decision is right or wrong, we must remember that everyone is right from his or her own perspective.

With the foregoing in mind, I present four major themes in the book you are holding. In the first is "today's crisis," in which I examine the problems we humans have saddled our forests with because we too often see them simply as unlimited commodities to be endlessly exploited for quick economic gain. The second theme is "tomorrow's hope," wherein I present a different way of perceiving the same forests, a way that will leave a forest legacy antithetical to the destructive one we are in the process of bequeathing our children into the foreseeable future and beyond. The third motif is the perpetuation of our public forests as a "biological living trust," a concept I use to discuss what I believe it will take to restore and maintain an ecologically sustainable forest for all generations.

By "our public forests," I mean those owned in common by you, me, and all the children, present *and* future, such as our national forests, national parks, and the forested lands held by the Bureau of Land Management, as well as those in the charge of the U.S. Fish and Wildlife Service in its system of refuges. "Our forests" extend from the pine forest of Florida, to the hardwood forest of Vermont, to the redwood forest of California, to the coastal rainforest of the Olympic Peninsula in Washington state and southeastern Alaska, to the spruce forests of interior Alaska. Like you, these forests have names, such as: Conecuh, Tongass, Apache, Coronado, Ozark, St. Francis, Modoc, Stanislaus, Roosevelt, San Juan, Choctawhatchee, Chattahoochee, Sawtooth, Hoosier, Daniel Boone, Hiawatha, Chippewa, Mark Twain, Lolo, Deschutes, Ochoco, Sumpter, Green Mountain, Chequamegon, and George Washington.

I have focused specifically on federal, public lands—those held in legal trust for *all* American citizens—to avoid getting into a debate over the *rights of private property*, which would create an unnecessary diversion from the purpose of this book. Nevertheless, if the forested landscapes of our nation are to be healed, the principles and concepts within these covers must be applied equally to public *and* private forestlands.

The fourth thematic thread is that we humans *need* to nurture something larger than ourselves if we are to become whole in the sense of reaching the spiritual peace of psychological maturity. That said, if we focus on restoring our forests—indeed our environment—to health, we will find there is no "problem" in the world that lies outside the realm of our thinking.

This being the case, as we nurture the forests toward ecological sustainability for all generations, we heal ourselves. As we heal ourselves, we heal our society—one person, one community at a time. As we heal society, we heal our environment. As we heal our environment, we begin to understand what it really means to be a compassionate human being entrusted with the care of planet Earth.

In order to fulfill our charge as caretakers of planet Earth, there are four additional things you must understand while reading this book. First, as we strive to maintain sustainable forests, we are faced with the constant struggle of accepting open-ended change and its accompanying uncertainties, which often give rise to fear of the future and a disaster mentality. We must therefore be gentle with one another and do whatever we do with love because there are no "enemies" "out there," only frightened people—some of whom abuse such social systems as politics and economics. Second, while ideas change the world, for that to happen, people must change their ideas. Put differently, people must change before ideas will change, and ideas must change before the world will change. Third, all we have in the world of real value as human beings is one another. If we lose sight of one another, we will find in the end that we have nothing of value after all. And fourth, this century is the pivotal one if we, the people, are to have any chance of healing our forests for those who follow us.

If the twenty-first century is to be the time of healing the forests of the world, we must be willing to work together as people who care about our own future and that of our children and our children's children. To heal the forests, we must be willing to share openly and freely any and all knowledge necessary to achieve that end. Moreover, we must be willing to cooperate with one another in a coordinated way, for cooperation without coordination is like a psychological ship without a rudder. We must therefore focus on healing our own forests and one another's while honoring one another's culture—even if we don't understand it.

Finally, I am telling it *as I see it.* I can do naught else because neither I nor anyone else *knows how it is.* While knowledge enables me to understand enduring elements that remain after my notion of the facts disappear, I still interpret what I see through the filters of my own perception. Therefore, I use the personal pronoun "I" when referring to my thoughts on a particular matter to identify the insights I have gleaned over a 64-year love affair with forests, especially those of western North America with which, and in which, I grew up. Be that as it may, I am well aware that others will come along to point out yet other insights that, in turn, will be a little more profound than mine and so elevate the consciousness of humanity a notch further in relationship to the forests of the world.

Due to the complexities of writing the text, there is of necessity a little overlap, and even periodic redundancies, within and among the parts of this book. Although I've focused on the Pacific Northwest for many of the examples because it's the area of much of my experience, the principles and concepts discussed in the text are generally applicable to all forests. It's thus critical that we both understand and accept the effects we cause by redesigning global forests, because we're simultaneously redesigning the structural and functional processes of the world, such as cycles in soil fertility; cycles in quality, quantity, and belowground storage of water; and cycles in climate.

Should you wonder why I am writing this book, the answer is simple. Today's decisions not only will determine the options of tomorrow but also will write the history of yesterday. Today, we have far more knowledge of the world in which we live than did our forebearers. We therefore have greater opportunities and responsibilities than they did because we are no longer an isolated continent but part of an interconnected global society, whether or not we fully understand the idea, whether or not we even like the idea.

Moreover, if humanity is to survive this century and beyond with any semblance of dignity and well-being, we must both understand and accept that we have a single ecosystem, with us sandwiched in the middle (the biosphere) between the upper atmosphere (air) and the lower lithosphere (the crust of the Earth, including the soil and water). And because this magnificent, living system—planet Earth—simultaneously produces all of the products of the world, including us, we would be wise to honor it and care for it. If we do not, if we cause too much damage to any one of the "spheres," we will be the authors of our own demise and that of all of the world's children.

Having suffered so much agony and grief as a young man watching the ancient forest I loved fall to the chainsaw, I cannot bear the thought of today's children going through what I did—especially when there is so little, so very little, ancient forest left. As I hike today, all I see are a few tattered patches of the seemingly endless forest I once knew. I feel in the deepest recesses of my being that we owe the children of today and tomorrow a better prospect for their future, which also means a better prospect for their children and beyond.

Just so you don't think everyone who is somehow connected with forestry is insensitive and intent on summarily cutting down all of the old trees, as many people imagine them to be, I once saw a District Ranger in the U.S. Forest Service show his tender emotions in public. This incident took place many years ago on the Umatilla National Forest, which straddles parts of southeastern Washington and northeastern Oregon.

I was facilitating the resolution of a long, ongoing conflict among a number of organizations, public agencies, Indian tribes, and industrial representatives, who were all vehemently fighting with one another over the management plan of the Umatilla National Forest. During one of the sessions, someone from an environmental group pointedly accused the District Ranger of being biased toward the timber industry. He, in turn, looked like he had just been shot in the gut with a dumb-dumb bullet, one with a cross cut into its nose so it flies apart on impact, which does incredible damage without immediately killing.

He turned to the person and, with quivering chin, misty eyes, and faltering voice, said, "Just because I'm a district ranger doesn't mean I have no feelings. I just had to sign-off on a timber sale that will clear-cut my favorite place in the world, and I'm sick at heart."

This book is a poignant one for me to write because I find myself immersed once again in the pain and grief of old feelings for that part of my soul that was torn away by the "cut and run, get it all" mentality of the timber industry as I was growing up. These feelings call to mind the translation of the Mayan word for human: "those who bear the burden of time." While time may be a burden I have had to bear, it's also a blessing because I have healed enough through the intervening years to write about the sorrows of the past and balance it with hope for today and tomorrow.

1. Tree in the middle is a western redcedar, note stringy looking bark. The tree on the right is a Douglas fir, note craggy bark, and the small tree on the left is a western hemlock, note smooth bark compared to the Douglas fir. (Photograph by Chris Maser and Larry D. Harris.)

2. Note the spire-like growth form of the subalpine fir in the right-hand corner of the photograph; Pacific silver fir is not quite so compact. (Photograph by Chris Maser.)

Chapter 1

Learning to See a Forest by Understanding its Dynamics

I would rather live in a world where my life is surrounded by mystery than live in a world so small that my mind could comprehend it.
— Henry Emerson Fosdick, American preacher and author

As a youth, I saw and felt the forest with my heart, and I found in it nurturance and safety. Today, I see the forest through the experience of years, training as a research scientist, and the heart of one in love with its song and mystery. As a youth, I felt its heart beat, today I marvel at the fluid dynamics of its pulsating cycles, the interplay of it parts, and the mystical union of its opposites—life and death, animate and inanimate.

As a youth in the 1950s, and even as a young man in the earliest years of the 1960s, I could stand on the shoulder of a mountain in the High Cascades of the Willamette, Deschutes, or Gifford Pinchot National Forests and gaze upon a land clothed in ancient forest as far as I could see into the blue haze of the distance in any direction. My sojourns along the trails of deer and elk were accompanied by the wind as it sang in the trees and by the joy of water bouncing along its rocky channels. At others times, the water gave voice to its deafening roar as it suddenly poured itself into space from dizzying heights, only to once again gather itself at the bottom of the precipice and continue its appointed journey to the sea—the mother of all waters.

Throughout those many springs and summers, the songs of wind and water were punctuated with the melodies of forest birds. Wilson warblers sang in the tops of ancient firs, while the plaintive trill of the varied thrush drifted down the mountainside, and the liquid notes of winter wrens came ever-so-gently from among the fallen monarchs as they lay decomposing through the centuries on the forest floor. From somewhere high above the canopy of trees came the scream of a golden eagle, and from deep within the forest there emanated the rapid, staccato drumming of a pileated woodpecker.

These were the sounds of my youth. This was the music that complemented the forest's abiding silence—a silence that archived the history of centuries and millennia as the forest grew and changed, like an unfinished mural painted with the novelty of infinite Creation.

But then something happened that forever changed my life. It was the 13th of October 1961, and I had just turned 23. The sky was a pure, cloudless blue. The snow atop Mount St. Helens glistened in the sun across Spirit Lake, which is in the Gifford Pinchot National Forest of Washington. I stopped eating the sun-ripened, sun-warmed black huckleberries and stood for a long time, a seeming eternity, on Grizzly Pass overlooking Spirit Lake. I had been in the mountains for three agonizing weeks, and in my heart I knew that I could never go back to the mountains of my youth. I would never again know the mountains as I had known them.

I had just spent three weeks struggling within myself, trying to figure out what I would do. I had seen virtually all of the places that I loved fall to the chainsaw. The last one I had just left—the valley of the Green River.

I still remember the trail along the river as it flowed through the ancient forest that was unbroken for 50 miles in any direction when I first knew it. The trail was dappled with soft green and bright yellow light filtered through a canopy of vine maples amid the towering trees. It was cool and deep, and the western red cedar trees that lay on the ground across the old trail were so big that I could not climb over them. I either had to find a way under them or go around them.

Deep in the forest, far from any human habitation, there was a soda spring, where the band-tailed pigeons liked to drink and from around whose edge they ate the ripe blue elderberries. As for me, I packed in some lemons and a little sugar and made the best lemonade from the bubbly water. I also made syrup from the elderberries, flavored it with wild ginger and salal berries, and poured it over my hot cakes and pan-bread. In addition, the river in those days had so many native cutthroat trout that I caught them by dangling a dry fly an inch above the surface of the water. Deer, elk, black bear, marten, mink, blue grouse, and an occasional ruffed grouse shared the river with me as I fished for my supper.

I had last camped there in the autumn of 1959, with no hint of the clear-cut logging that had for years been working its way up the river. But this time (1961), as I rounded the last bend in the trail just before the soda spring, I was struck dumb and sick at heart by the view. As far as I could see there was one giant clear-cut with human garbage scattered along the miles of logging roads. I caught one fish—and it had a hook in its mouth.

That was the day I learned the depth to which grief could plunge my soul. From that day to this, I know the endless grief of seeing one more clear-cut, one more mountainside denuded, one more roadless area dissected and fragmented, one more landscape becoming a homogenous patchwork of monocultural plantations of trees.

On that day in 1961, standing on Grizzly Pass looking down on Spirit Lake, I knew I faced the crossroads of my life. I knew I had to make a choice: flee into the wilderness of Alaska or Canada or face the cities I despised and fight for the survival of the forests I loved so much. I don't know how long I stood there, but the sun was setting over Mount St. Helens, and I still had miles to go. Like the Native youths who had gone before me into the sacred heart of the ancient forest, I left behind my youth. Unlike those youths of old, however, I also left my heart in the care of the land, for where I had to travel there were no sacred forests left. From the day I watched the sun set over Mount St. Helens in 1961 to this, I have spent the better part of my life as a scientist working to help society understand and accept the ecological and social value of diverse forests in all their stages of development—forests that are inseparable from and vital to the well-being of humanity through all generations.

The Novelty of Infinite Creation

While my youth has fled with the passing years, today I behold an even greater wonder as I once again hike the high mountain trails. Now I see the infinite, creative novelty of the forest's three basic components: composition, structure, and function as they are detailed in the trees.

For simplicity's sake in helping you to understand what I see, I will use five coniferous (evergreen, needle-bearing) trees—Douglas fir, silver fir, western hemlock, western red cedar, and Engelmann spruce—to illustrate the dynamic interplay of composition, structure, and function. There are, however, a few other coniferous trees in the forest besides these five, as well as many deciduous (non-evergreen, leaf-bearing) shrubs and a few deciduous trees of various kinds, which says nothing of the numerous varieties of herbaceous (non-woody) plants that grow relatively close to the ground.

Composition consists of the number and kinds of plants that grow in a particular area and the length of time they live (*photo 1, page 18*). The longevity of a particular kind of plant is critical because that is the length of time the plant affects the site on which it grows:

- Douglas fir lives more than 750 years, has a maximum life span of about 1,200 years, and might lie another 500 years as it decomposes on the forest floor.
- Silver fir lives more than 400 years and has a maximum life span of about 600 years.
- Western hemlock lives more than 400 years and has a maximum life span of about 500 years.
- Western red cedar lives more than 1,000 years and has a maximum life span of about 1,200 years.
- Engelmann spruce lives more than 400 years and has a maximum life span of somewhat more than 500 years.

Structure, on the other hand, is an outcome of the composition of plants that grow in an area because each individual plant and each kind of plant grows differently. The cumulative effect of how they grow creates the vegetative structure of the area:

- Douglas fir grows to a maximum of about 260 feet tall; has a straight, wood stem with a maximum diameter of about 175 inches; rough, craggy bark; and relatively stiff branches well clothed in moderately stiff needles.
- Silver fir grows to a maximum of about 180 feet tall; has a straight, wood stem with a maximum diameter of about 80 inches; smooth bark with bumps containing pitch; and stiff, rather short branches, giving the tree a tapered, spire-like appearance (*Photo 2, page 18*).
- Western hemlock grows to roughly 200 feet tall; has a straight, wood stem with a maximum diameter of about 100 inches; rather smooth, finely textured bark *(photo 1, page 18)*; and lacy branches with rather short, sparse needles.
- Western red cedar grows to more than 195 feet tall; has a straight, wood stem with a maximum diameter of about 100 inches; rather smooth, stringy bark *(photo 1, page 18,)*; and droopy branches.
- Engelmann spruce grows to approximately 160 feet tall; has a straight, wood stem with a maximum diameter of about 90 inches; scaly bark; and relatively droopy branches well clothed in stiff needles.[1]

The combined features of composition and structure allows certain *functions* to take place within a given area of the forest:

- Douglas fir's rough, craggy bark offers numerous crevices in which small bats, such as the little brown bat, sleep during the day, whereas the fir's stiff, well-clothed branches make excellent sites for a variety of birds to nest. Old firs also support generations of red tree mice[2] within a single tree (*photo 3*). The mice not only dine on the fir's needles but also lick the dew and rain off them. As well, the old firs are used for roosting and nesting by northern spotted owls (*photo 4, page 25*), the tree mouse's main predator.
- Silver fir's tapered, spire-like shape of stiff, rather short branches, while not particularly conducive for nesting by a wide variety of forest birds, readily withstands wind and easily sheds the snows of winter.
- Western hemlock's rather smooth, finely textured bark and lacy branches hold little value as wildlife habitat, but the tree's relatively short life span creates a fairly steady supply of large snags (standing, dead trees, for cavity-nesting birds, such as pileated woodpeckers, and cavity-nesting mammals, such as northern flying squirrels, which appropriate the abandoned cavities of the pileated woodpecker. Snags with loose, but attached, slabs of bark offer roosting sites for silver-haired bats.
- Western redcedars often have much-enlarged stems at ground level that frequently become hollow in old trees due to rot. These hollows are ideal winter dens for hibernating black bears, especially for female bears because they give birth to their cubs during hibernation. In addition, large western red cedar trees often become hollow snags because of a long infestation of heart rot. These hollow snags are critical nesting habitat

for Vaux's swift, a small bird that flies into the hollow snag from the top and fastens its nest to the inside wall of the dead tree. While the droopy branches have little value for nesting birds, northern flying squirrels use the stringy bark to construct some of their nests.

• Engelmann spruce's relatively droopy branches are, nevertheless, well-enough covered in stiff needles that some birds use them for nesting.

Which of these trees and their accompanying vegetation happen to grow in a particular area of the forest is determined by such things as elevation, aspect of the slope, proximity to water, and shade. By way of illustration, let's take a brief look at these four factors.

Elevation plays a major role in what kind of forest can grow where. In western Oregon, for example, the forests below 4,500 feet in elevation along the western slopes of the Cascade Mountains are composed primarily of Douglas fir, western hemlock, grand fir, and western red cedar. Above 4,500 feet along the western slopes of these same mountains, the forest consists of Douglas fir, western hemlock, grand fir, and western red cedar, all of which give way to subalpine fir, silver fir, noble fir, mountain hemlock, Engelmann spruce, Alaska yellow-cedar, and white-bark pine as the elevation increases.

The role played by the *aspect of a slope* is perhaps illustrated best and most simply along the western edge of the Willamette Valley of western Oregon, where Douglas fir grows on the moist, north-facing slopes of the low hills, and Oregon

3. Old-growth Douglas fir, note craggy bark. (USDA Forest Service Photograph by Jerry F. Franklin.)

white oak grows on the dry, south-facing slopes of the same hills. As well, there is a remarkably sharp line of demarcation between the two plant communities that go up the west side of a hill, over the top, and down the east side of the hill, effectively juxtaposing the fir and oak.

With respect to coniferous trees, *proximity to water* is perhaps most pronounced in western red cedar, which grows well in relatively wet sites, although Douglas fir, western hemlock, and grand fir also grow adjacent to such bodies of water as streams and rivers. With respect to deciduous trees, the banks of streams and rivers (termed "riparian zones") are dominated by red alder and big-leafed maple below approximately 1,800 feet in elevation and by generally by conifers above 1,800 feet. If you happen to visit the Pine Barrens in New Jersey, on the other hand, you will find Atlantic white cedar along the waterways and pitch pine on the drier sites.

How well a species of plant tolerates *shade* is another factor that dictates where it can grow. Douglas fir requires a goodly amount of sunlight and does not grow well—or at all—in areas of dense shade. In contrast, western hemlock and western red cedar grow well in dense shade, but do even better in light shade and full sunlight. Contraposed, grand fir is more shade tolerant than Douglas fir, but not quite as shade tolerant as western hemlock and western red cedar.

While elevation, aspect of the slope, proximity to water, and the relative tolerance of shade are important, a myriad other factors are also involved in what kind of forest grows where, and fire is clearly one of the principle determinants in the coniferous forests of North America. Fire not only adds to the overall diversity of habitats within the forest but also "fire proofs" it at the same time by periodically removing the easily combustible fuels on the ground.

Here, it's critical to understand that a forest stores its excess energy in dead wood and must periodically dissipate that energy if the forest is to survive long enough to become truly old. Unless the dead wood is periodically removed through low-intensity, ground fires, the energy stored in the dead wood continues to accumulate (compounding the potential intensity of the next fire) until, when a fire comes, it burns so intensely that it kills the forest and starts it over. Although the suppression of forest fires began well before I was born, the ancient forests of my youth seemed little affected by it, despite the fire lookouts that squatted like little forts on various high points in the rough terrain.

The giant Douglas firs that I traveled among were seldom injured by these low-intensity fires as they crept around on the forest floor because the fir's outer bark was ten to twenty inches thick at their bases, which protected their living tissue from the heat. Additionally, the lowest branches of their crowns were often sixty feet or more above the flames, far too high to sustain damage, much less allow the flames to reach the tree's crown, where it could explode into a raging inferno.

The reason an old fir's lowest branches were usually so high above the ground stems from the fact that most of their regeneration following a fire was in the form of dense thickets that prevented the penetration of light. This situation comes about because the paucity of light progressively kills (prunes) the lowest limbs, which ultimately rot and fall off as the trees grow. Due to competition for light, water, and nutriments, most of the mortality in such a thicket of young trees is caused by the continual suppression of the weakest by the stronger ones. This kind of mortality can be thought of as Nature's way of selectively "pruning" the trees most susceptible to fire. (*photo 5*)

By the time the trees reach 120 years of age, however, a thicket is open enough that mortality is on a single-tree basis. Moreover, the remaining trees begin to acquire much more individualistic characteristics because they have greater freedom to grow. As these trees matured, the bark on the downhill side of

4. Spotted Owl. (USDA Forest Service Photograph by Eric Forsman.)

5. Young forest under the influence of "suppression." Note how the lack of light causes the lower branches to die, rot, and fall off and the weak trees to die and fall, adding to the stored energy that litters the forest floor. (USDA Forest Service Photograph.)

their stems becomes thicker and harder than that on the uphill side. Because fire usually proceeds uphill, and because the bark of ancient firs is thickest, hardest, and most resistant to fire on the downhill side, they are further protected from the flames of low-intensity fires.

In a sense, ponderosa pine, which grows along the eastern flank of the High Cascade Mountains, is even more resistant to fire than Douglas fir. Like old Douglas firs, old ponderosa pines have their first branches high above the ground and have thick bark at their bases. Unlike the fir, however, ponderosa pine has bark that is not only plate-like in structure and so slow to burn but also has little pustules of pitch scattered over the outer surface of the bark. If the bark does catch on fire, the pustules explode and literally blow the bits of burning bark off the tree.

In a few areas of Oregon, where fires did not occur over many centuries, such as the coastal portion of Tillamook County in the northwestern part of the state, western hemlock replaced Douglas fir because hemlock's offspring can grow in the dense shade of the parent and grandparent trees, whereas Douglas fir cannot. Hemlock's ability to be self-perpetuating makes it a "climax" forest. ("Climax" is the culminating stage in the succession of a forest, in which the vegetation is self-reproducing and thus has reached a relatively stable condition through time.)

Douglas fir, on the other hand, is considered a "fire subclimax" forest because it requires the periodic return of fire to clear away the shade-tolerant—but fire-susceptible—western hemlock and western red cedar that would take over dominion of the area without the presence of fire. ("Subclimax" is the persistent succession stage immediately preceding the climax stage because of the effects of fire or other major, recurring disturbances.)

When a forest-replacing fire does occur, it seldom kills all the trees. Various numbers of live trees are left standing as individuals, as small island-like clumps, or in "rows" commonly termed "stringers." Most of the trees that are killed by the flames and heat remain standing as snags through subsequent decades. The burned forest then commences what is termed "autogenic" (self-generating) succession. Autogenic succession can be characterized by "successional stages," a concept that refers to the characteristic developmental stages that a forest goes through from bare ground to an ancient or "old-growth" forest.

Autogenic succession works as follows: Grasses and other herbaceous plants are the first to grow in a burned area and so constitute the first successional stage following a fire. In their occupation of a given site and in their growing, they gradually alter the characteristics of the soil, such as pH, until it is no longer optimum for their survival and growth. Their offspring may germinate but not survive, creating areas wherein only the parent plants remain. As the offspring succumb to the changes in the soil and the parent plants age, die, and are not replaced, openings appear in the vegetative cover that allow shrubs to become established.

In this way, the herbaceous stage changes to an early shrub stage (*photo 6*).

6. Note shrub stage in immediate foreground, old-growth Douglas fir forest and snags just behind shrubs, mature forest just beyond the old-growth, young forest in upper right-hand corner and left center, and the recent clear-cut (right down into the stream bottom) in the upper left-hand corner. (Photograph by Chris Maser.)

In turn, the shrub stage follows the same process as the herbaceous stage and gives way to tree seedlings that, in combination, become the shrub-seedling stage. As the shrubs die, the shrub-seedling stage gives way to the sapling stage, which grows into a young forest stage, a mature forest stage, and finally an old-growth forest stage—that is, until succession starts over. The six generalized autogenic, successional stages that a western coniferous forest goes through can be characterized thusly: herbaceous —> shrub-seedling —> sapling —> young forest —> mature forest —> old-growth forest —> *fire* or other disturbance, which starts the cycle over, and eventually over again.

There are, however, two caveats to this mode of succession that need to be mentioned here. The first counters the classical, forestry-textbook notion of *discrete* "successional stages" that replace one another in a predetermined, orderly fashion through time. Rather than "discrete" stages of development that precede in an orderly succession, one after another, the "stages" of forest development form a complex continuum wherein each stage builds on the dynamics and ecological nuances of the preceding one. Hence, no two forested areas develop alike in an absolutely repetitive fashion.

The second caveat is one in which agriculturally oriented foresters (those who see trees as analogs to agricultural crops, as opposed to the definable expression of a forest) short-circuit autogenic succession by planting trees whenever they wish—invariably out of sync, and thus out of harmony, with Nature's cycles—and then define any unwanted vegetation as "competition" with their "crop trees." Such "competition" is *management-created* by attempting to force Nature's time frame into conformance with short-term economic ends. This shift in focus comes about when a forester "cannot see the forest for the trees." Put differently, to a linear-minded forester, evenly spaced trees grown like an agricultural crop become synonymous with Nature's forest.

That notwithstanding, the "stages" of autogenic succession are a dynamic web of interrelated events in which none of the properties of any part of the web are fundamental; rather, their mutual interrelationships determine the structure of the whole. For example, a viable seed must exist before a seedling can exist; a seedling must exist before a young tree can exist; a young tree must exist before a mature tree can exist; a mature tree must exist before an old tree can exist; and a large, live, old tree must exist before a large snag or large fallen tree (often thought of as a "log," *see photo 3, page 23*) can exist. In other words, a large snag and a large fallen tree are only altered states of a large live tree. And it all begins with a seed that is allowed to grow and perform its appointed functions in the forest.

The crux of maintaining diversity within a forest is that things must exist before they can be in relation to one another, part of the unity of all things. This specifically includes the diversity of processes—such as fire—within a forest because diversity (compositional, structural, and functional) either maintains or alters the speed and direction of succession and hence the resulting plant community. Therefore, a forest has neither a single state of equilibrium nor a single deterministic pattern of recovery.

When thinking about landscapes in the Pacific Northwest, I am often reminded of the fires, both large and small, that over the millennia shaped the great forests I knew as a youth. What seems clear to me now is that Nature's cycles (such as autogenic succession) are not perfect circles, as they so often are depicted in the scientific literature and textbooks. Rather, they are a coming together in time and space at a specific point, where one "end" of a cycle approximates— but *only approximates*—its "beginning" in a particular place. Between its beginning and its ending, a cycle can have any configuration of cosmic happenstance.

In this sense, Nature's ecological cycles can be likened to a coiled spring in so far as every coil approximates the curvature of its neighbor but always on a different spatial level (temporal level in Nature), thus never touching. The size and relative flexibility of a spring determines how closely one coil approaches another. Regardless of the size and flexibility of the spring, its coils are forever reaching outward.

With respect to Nature's ecological cycles, they are forever reaching toward the novelty of the next level in the creative process and so perpetually embracing the uncertainty of future conditions. In thinking about the great forests I used to know, and those parts through which I can still hike, I am awed by all of the factors that must come together to create a particular place as I perceived it or remember it, not just the events themselves but also the cycles in which the events are embedded.

Besides the cycles of fire, such things as age, disease, injury, and wind also played significant roles in the forests of old. Trees that died of disease and fell often seemed to be randomly oriented on a slope, whereas those blown over by wind usually fell in a relatively consistent direction. The fallen trees become mixed with snags that broke as they fell and other trees that periodically toppled because their roots were weakened by fungal rot. Regardless of how jumbled the fallen trees were at first, they all eventually became part of the soil from which the forests of today grow.

Compared to times past, few large, stable, fallen trees remain along the contour of a slope to help reduce erosion by forming a barrier to creeping and raveling soil that gradually works its way downhill to the beck-and-call of gravity. If the soil's travels are not interrupted, it eventually ends up at the bottom of the slope. On the other hand, soil deposited along the uphill side of fallen trees (*photo* 7) reduces the loss of nutrients from the site and forms excellent places for the establishment and growth of vegetation, including seedling trees, such as those of western hemlock.

As vegetation becomes established, it helps to stabilize this "new soil." When invertebrates, such as earthworms, and small mammals, such as red-backed voles, commence burrowing into the new soil, they not only enrich it nutritionally with their feces and urine but also constantly mix these nutrients into the soil through

7. Note soil that has collected on the uphill side of this old log. (USDA Forest Service Photograph by James M. Trappe.)

their burrowing activities.

The interactions of fallen trees with soil are mediated by steepness of slope and ruggedness of terrain. A tree on flat ground is much more likely to be in contact with the soil along its entire length than one on steep or rough ground. The proportion of a tree in contact with the soil determines the water-holding capacity of the wood. In a dense forest, moisture retention in the wood during summer drought is greatest on the side of the tree in contact with the soil. In turn, the moisture-holding capacity of the wood affects its internal processes and, therefore, how plants and animals use the fallen tree as habitat. By this I mean the more saturated the wood is with water, the slower it decomposes and the fewer organisms can live in it because the moisture limits the amount of available oxygen in the wood. How a tree lies on the forest floor and the duration of sunlight it receives also have a strong impact on its internal processes and biotic community, again because the damper the site, the wetter the wood, the less the available oxygen.

Even today, when you look closely at the surface of the floor in an old forest that has not been disturbed by human endeavors, it becomes apparent there is no such thing as a smooth slope. The forest floor is roughened by the scattered stumps, pieces of collapsed snags, and whole fallen trees, their uprooted butts, and the pits and mounds left by their uprooting. Living trees roughen the surface of the forest floor by sending roots outward along slopes, often near the surface. And tree trunks distort the surface by sloughing bark and by arresting creeping soil at their bases.

Decomposing woody roots of tree stumps also have distinct functions in the forest. Tree roots contribute to the shear strength of the soil, which is a root's ability to hold soil in place without being pulled apart by gravity. Declining shear strength of decomposing woody roots increases mass soil movement (which you might think of as a "landslide") after such disturbances as catastrophic fire—or today, road building and clear-cut logging. Another related function of decomposing tree stumps and roots is the frequent formation of interconnected, surface-to-bedrock channels that rapidly drain water deep into the soil from heavy rains and melting snow. The collapse and plugging of these channels as roots decay can force more water to drain through the soil matrix, reducing soil cohesion, and increasing hydraulic pressure that, in turn, can cause mass soil movement. Finally, decaying woody roots that harbor root-rotting fungi, such as the laminate root rot fungus, contribute to the continuation of the disease in the forest as the roots of healthy trees come into contact with the already-infected dead roots.[3]

And so the forest of my youth continued to change, a mural of infinite novelty detailed in my book *Forest Primeval: The Natural History of an Ancient Forest*.[4] There was, however, an unsettling part of the change.

When Silence was Replaced by Grief

By the early 1960s, I began to notice a different kind of change, one that

filled my heart with dread. The warm summer winds, coming up from the valleys, carried with them the faintest of roars. Chainsaws were coming. That meant
only one thing—devastation to the ancient forests I grew up loving.

With their loud, smoking engines that rend the forest and deafen the ear,
chainsaws stifle the songs of birds far and wide as metal teeth speed through the
wood too fast for anyone to really get to know the tree they're cutting. The stench
of gasoline and oil fouls the air and permeates the sawyer's clothing. And still
today, men with chainsaws cut trees so fast they fall like jackstraws as forests are
leveled around the world.

Although the great American forest, chronicled by Rutherford Platt,[5] had
been progressively logged for decades, it was different in that men got to know
the trees they were cutting. The chainsaw, the harbinger of increasingly mechanized logging, swiftly removed the logger from the once-intimate relationship of
working with a tree within the context of its forest—of a human getting to know,
really know, something about the tree he was felling and bucking to be converted
into lumber or simply cutting for firewood. An example of the once-intimate relationship with the tree that a man was cutting is eloquently memorialized in Aldo
Leopold's *A Sand County Almanac* as he cut an ancient oak with a crosscut saw.
Cutting through the oak, Leopold stopped periodically to rest, and in so doing
recalled episodes in Wisconsin history through which the tree had lived. In the
process, he became acquainted with the tree, both in cutting it for firewood and
in burning it to warm his cabin.[6]

Having used a crosscut saw (or "misery whip," as I knew it in decades past)
to cut firewood, I am well acquainted with the exertion of pulling sharp, metal
teeth through resistant wood and feeling the growing tiredness of muscles as
they begin to rebel at the seemingly unending repetitive motion. But as the saw's
teeth severed the wood, I could still hear the songs of winter wrens, smell the
scent of freshly cut wood in the clean autumn air, and feel the quietude of the
forest surrounding me as giant Douglas firs, western hemlocks, and western red
cedars towered overhead.

What the arrival of the chainsaw meant to the forests I loved really came
home to me as I studied the red tree mice, whose populations were constantly
decimated by clear-cut logging. The destruction of their habitat was particularly
painful to me because I had grown to love the little red mice so very much. In
fact, each time my research necessitated killing one of them, I'd lock myself in my
office and cry almost uncontrollably during and after the whole process. Once
again, the unrelenting grief of continual loss knocked at the door of my heart.

I know grief is a critical emotional process through which we mediate our
acceptance of a painful circumstance and reshape ourselves in relationship to a
new reality. To be healthy, we not only must be allowed to grieve but also must be
given permission to grieve, especially in our death-denying society.

People have long used rituals to help themselves and one another mourn
and recover from grief. Funerals and memorial services function as a rite of passage between the initial shock of loss and the longer, more private and difficult

phases of grieving. Most customs of contemporary mourning are directed at the acute loss of a person or pet during the first weeks and months of the grieving process. But our environmental and social losses are intermittent, chronic, cumulative, and without obvious beginnings and endings.

With the death of a person or animal we love, there is a finality beyond which healing begins. With our environment, however, there is no such finality—only a sense of increasing environmental stress wrought by our exploding human population, which is coupled with a sense of accelerating change and loss. How do we deal with this ongoing, underlying feeling of emotional distress that often borders on anguish?

Those of us trained to deal primarily through our intellect are too often cut off from our feelings and so try, as best we can, to stoically minimize the pain. We wax philosophical in our acceptance of the "inevitable" deforestation of our favorite neighborhood hill for that new housing subdivision or the razing of the historic opera house for that new shopping mall. On the other hand, those who are openly in touch with their feelings and have the courage to express them quickly find themselves in a hostile environment, where they are accused by monied interests of caring more for a neighborhood hill with its trees, wild flowers, and butterflies or for a spent opera house than for people in need of housing or shopping malls.

It is no surprise that both our internal and external worlds often make grieving for the loss of our environment a most difficult and uncertain process. The personal danger of the grieving process is compounded not only because we are so vulnerable in our despair but also because we must defend, without respite, our deepest personal feelings against the cold insensitivity of materialism.

Is it any wonder that we resist change, that we are committed to protecting our existing values, representing as they do the safety of past knowledge in which there are no unwelcome surprises? Is it any wonder that we try to take our safe past and project it into an unknown future by skipping the present, which represents change and holds uncertainty, potential danger, and grief?

We have almost no social support for expressing grief over environmental losses, as I was to learn in graduate school, where stoic scientists are trained. Honest conversations about grief, which have the potential to occur at the bedside of a dying loved one, are dangerous at a conference table, where the issue is land-use development and someone's interpretation of their sacred rights of private property and their profit margin.

Nevertheless, we deem ourselves a free democratic society, one in which *any* issue is fair to discuss in a public forum. Here, in public debate, is the open forum wherein we can express our feelings. This is important because we are *not* required in our democracy to separate our feelings from our thoughts concerning a topic. All that is required is the courage to come together and act, the knowledge to know how to act, and the vision to know where to go. Yet, today, expressions of emotions and grief are being curtailed in public debate in the United States.

The freedom and survival of democracy, as perhaps one of humanity's most noble achievements, is based on accepting change as an ongoing process to be embraced, not as a condition to be avoided at any cost. To embrace change, however, *does not mean* blindly casting out the old as we unquestioningly clutch the new, especially when that which we are being forced to give up is finite in supply, intangible in value, and irreplaceable once lost. Furthermore, while the choices belong to us, the adults of today, the consequences belong to the children of today and all the tomorrows.

Public debate is the only forum we have to help ourselves and one another consciously integrate proposed environmental changes into our lives and in the process grieve for the imminent loss of a safe and known past as we step into an unknown and uncertain future. Open public debate must be protected at any cost if we are to retain a free democracy for the future generations. While I was not astute enough in graduate school to understand all this, I was keenly aware of my grief, and went to the only person with whom I could discuss it—Dr. Kenneth L. Gordon.

Ken, the professor in charge of my graduate studies, was an exceedingly gentle man of tremendous artistic talent in writing, drawing, woodcarving, and photography, in addition to which he exhibited boundless creativity in his pursuit of natural history. Moreover, he, too, suffered from the same grief that I did. But it was his philosophy of minimal, ethical disturbance of Nature, his gentleness with all things living, and his deep regard for the living spirit he saw in all things that still influences me, an influence that began with my barging into his office one day in a blind rage over the destruction of a forest where I was studying red tree mice.

Telling me to close his office door, he motioned me to sit while he tapped the spent tobacco out of his ever-present pipe, refilled and lighted it. He then regarded me for a long, silent moment, as though making up his mind about something. "Chris," he said, "you'll find in life that most people are so busy with their day-to-day affairs they scarcely notice anything else. I came here to Oregon in 1926. It was much more peaceful then with far fewer people. In fact, I think I drove every road in the state within a decade. I've seen a lot of destruction of the state within these past thirty-seven years [it was now early 1963]. That's why I started photographing what I call 'The Passing Scene.' It's my way of remembering the part of my life that's fading into history, never to return. Creating an archive of visual memories is my way of softening my grief."

"Doc," I interrupted his slow, methodical speech because—as impolite as it was—experience had taught me that Ken was easily sidetracked into examining hitherto unexplored, mental rabbit trails, "what are you getting at?"

"Well," he continued, "like me, you'll see much that you love disappear in your lifetime before this modern notion of 'progress.' And, like me in my youth, you'll feel powerless to stop it. In fact, I came west when I graduated from Cornell with my Ph.D. because I loved the open country. I wanted to get away from the destruction I saw on the East Coast. But I couldn't get away from it. It's here too!"

"How did you even know about the openness of the West if you were at

Cornell?"

"Because I was raised in Fort Collins, Colorado. Anyway, I've found that anger and force net only despair. You've got to examine your feelings, and then go through them to reach the rational logic that will tell you what to do."

"How," I asked impatiently, "do I know when I've reached this 'rational logic' of which you speak?"

Ken went to the small blackboard on his office wall, picked up a piece of chalk, and drew the simple depiction of stairs. Pointing to the top step, he drew a horizontal line outward from it and said: "This was Oregon as I saw it in the late twenties and early thirties, and to me that was the optimum."

"Okay," I ventured, "but I was born in 1938 and gained my perception of Oregon at its optimum in my teens and early twenties as I wandered the trails of the Coast and Cascade Mountains, as well as the high-desert steppe east of the Cascades. But now I find most of the places I love already falling to the roar of chainsaws."

Ken regarded me for an instant. Then went down two steps to 1950 and drew another horizontal line: "This is when your perception of Oregon began to emerge." Drawing a third horizontal line outward from another step down (1960), he continued, "This is when you really began to notice the changes taking place. It is part of the human condition to notice change at intervals. If you love Nature and the natural world around you, then you have the sensation of descending the stairs from top to bottom, with each step a loss of something cherished. But, if your interest in life is tied to something like technology, then you ascend the stairs with each new invention, such as the chainsaw. In the end, the desirability of a given change is based on one's perception of the circumstance that precipitated the change."

"Ken, this is clear," I said with some frustration in my voice, "but how will I know when I've reached this 'rational logic' you speak of?"

"You'll feel it; it's an intuitive sense of harmony with what's right, with what will work. For instance, did you ever consider that most people don't even know the consequences they cause in and to Nature simply because they're uninformed. They're not bad people; they're just ignorant. Not stupid—ignorant! Remember, scientists tend to write for and speak to one another; they've scant interest in informing the public. Perhaps that is something you could do—educate the public."

"How?"

"When the time's right, you'll know what to do and how to do it. But first, you have to learn the basics. You have to pay your dues before you can speak. So, back to your studies!"

Ken always saw something to which others were blind. He saw the spiritual laws that lay behind all science and technology, the spiritual laws that underpin the Universe. And in his quiet way, he planted in my mind, heart, and soul the seeds of his vision that I might one day see what he saw, a truly spiritual world.

Following that 1963 visit with Ken and my graduation from college, I came across a couple of quotations that recalled Ken's counsel from the shadowlands of my memory: "When the time's right, you'll know what to do and how to do it." The first comment was penned by Thomas Jefferson in a letter to William Charles Jarvis on September 28, 1820: "I know no safe depository of the ultimate powers of the society but the people themselves, and if we think them not enlightened enough to exercise their control with wholesome discretion, the remedy is not to take it from them, but to inform their discretion by education."

"Enlighten the people generally," wrote Jefferson, "and tyranny and oppression of body and mind will vanish like evil spirits at the dawn of day." The second remark came from British economist E. F. Schumacher who, while sharing Jefferson's notion about education, also saw its folly in contemporary industrial societies: "If Western civilization is in a state of permanent crisis, it is not far-fetched to suggest that there may be something wrong with its education. More education can help us only if it produces more wisdom."

Because knowledge equates to the power of choice and access to knowledge equates to empowerment, I have consciously heeded Ken's counsel. In keeping with that counsel, I have often written to "enlighten" the "people" about forests and forestry, although I am not, by training, a forester. Trained as a vertebrate zoologist and natural historian, I got into forestry because, years ago, when I asked foresters biological questions about the effects forestry practices had on the animals I was studying, all I got were economic answers that had little or nothing to do with the biology of forests. So I came to understand that the profession and practice of forestry had been largely reduced to exploitive forestry. By exploitive forestry, I mean the practice of forestry that abuses both the forest *and* the land in the human drive for immediate profits without regard for the negative consequences passed forward to the next generation. In this sense, exploitive forestry is the most economically efficient and repeatable commodity extraction attainable in which the conversion potential of trees into products, such as lumber, is realized in the shortest possible time—in essence, use it or lose it.[7]

Strict, short-term, bottom-line, exploitive forestry continues apace the world over (*photo 8, page 36*).[8] It is, therefore, with the generations of children in mind that I chose the title of this book to reflect the gift of unconditional love I think we, the adult guardians in and of the present, owe all generations to follow, if we are to legitimately consider ourselves sentient beings. After all, building a community for happy children begins with maintaining a biologically sustainable forest to supply the community with such things as clean water. Thomas Jefferson, because of the environmental destruction he saw, stated the same idea this way: "Then I say the Earth belongs to each generation during its course, fully and in its own right; no generation can contract debts greater than may be [or can be] paid during the course of its own existence."

Today, many people would refer to Jefferson's statement as "sugar-coated idealism" and quickly contend that greed is the fundamental current of human nature and so floods our social psyche that we are incapable of raising our consciousness to a higher, more benevolent level. I thoroughly disagree!

Study of living hunter-gatherer societies allows an excellent case to be made that human survival is both directly and concretely dependent on mutually sustaining physical and psychological cooperation and coordination in life's daily endeavors.[9] In contrast, capitalism—the socially negotiated construct envisioned by a few intellectuals of the Industrial Revolution—is both the seedbed and nursery of our Western-style "greed." Moreover, the grip of capitalism advanced so gradually within the realm of human awareness that it was hardly noticed before the monied few who focused on the power of control were able to co-opt the livelihoods and destinies of the masses. The insidious nature with which capitalism has enslaved humanity in destructive competition is reminiscent of a frog placed in a container of water that is heated so slowly the frog is cooked to death without realizing it.

With the cooking frog in mind, I posit that capitalism, as an ensconced, social institution, is more fundamental to our current forestry problems than is human nature in general. If this is correct, there is hope for the ecological sustainability of our public forests since people have the ability to raise the level of their consciousness and change their behavior simply because they choose to.

To this end, with humility, I offer some of what I have learned about the reciprocity of our human relationship with the temperate coniferous forest. By forest, I mean a "natural forest," a forested area that still contains most of the principal characteristics of the indigenous ecosystem. I am, in this sense, describing my understanding of a forest, an understanding that to me is an expression of

8. Exploitive forestry in the Pacific Northwest; note Nature's mature Douglas fir forest on the right and the timber industry's clear-cut in the center. (Photograph by Chris Maser.)

the interactions among time, soil, biota, climate, and place. I am also describing my understanding of sustainable forestry after having spent many years as a scientist trying to figure out what I mean when I use the words "forest" and "sustainable." In the end, I must conclude that each reader will decide for herself or himself what a forest is and what sustainable means. And that is as it needs to be in an ever-evolving world.

What I am going to say is aimed specifically at public lands under the jurisdiction of the federal government: (1) Forest Service—our national forests and national grasslands, (2) Bureau of Land Management—some forests, but mostly rangelands across much of the West, (3) Park Service—national parks and national monuments, and (4) Fish and Wildlife Service—wildlife refuges (*see maps on pages 39-41. The legend is on page 41*)). I have focused on these public lands for two reasons: (1) they cover much of the American West—Oregon, for example, is about 50 percent public lands and Nevada is about 85 percent public lands, etc. and (2) as I said in the Preface, they *are* held in legal trust for American citizens *in* all generations and thus not subject to the "rights of private property."

That said, the ecological principles and concepts I will address apply also to private lands. Only if the principles and concepts are applied to all forests—public *and* private—will our national landscapes be healed from decades of extractive abuse. Only if the principles and concepts are applied to all forests—public *and* private—will we, the adults of today, pass to future generations an unconditional gift of social-environmental sustainability that is free from ecological liens and encumbrances, as a true gift must be.[10] Anything less is an ecological debt passed forward to the next generation for payment.

Yet, the children are still incurring a compounding ecological debt through exploitive forestry, which is still being taught in our universities, despite all the scientific knowledge pointing to its ecologically destructive practices, as well attested by Nate Wilson, a 2002 graduate of the forestry program at the University of Georgia:

> Today . . . the vision of a forester most often espoused in class resembles an investor, trying to extract the maximum return in the shortest time. I have been taught how to characterize the forest as a commodity, how to quantify it in dollars and tons. I have learned what silvicultural techniques should be used to achieve the shortest rotation lengths. . . . I can discount costs and revenues and calculate bare land value for an infinite number of rotations. But I haven't learned how to actually perpetuate the forest for an infinite number of rotations, how to ensure that the living systems remain healthy.[11]

This book is a synthesis of ideas about social-environmental harmony and sustainability that I have been struggling to understand for more than thirty years, both as a human being and as a scientist. It has been one of my more important struggles because I am living in the middle of an ongoing cultural revolution in which society is trying to find the essence of what it means to be human, a search that expresses itself through changing human values. Although culture is not genetically inherited, it can be learned from the past, modified in the present, and passed on to future generations as a "story line" guiding the existence of a people. The notion of culture poses two questions: (1) What happens when a shift in cultural values in one part of a society rips apart the entire social fabric? and (2) How do we heal the social rupture that results from such a shift in cultural values?

Trying to answer these questions helps me with my understanding of the forestry profession, one that is relatively young, yet rich in experience and noble in vision. Sadly, the vision of forestry—once on the cutting edge of scientific knowledge and social responsibility—has dimmed and is rapidly being relegated to cultural history. I say this because listening to the old foresters I knew aupervisor of the Siuslaw National Forest) gave me a sense of their deep caring for the forest *as a forest*. And they *knew* the forest in those early years of the teens (when Fred began his Forest Service career), twenties, and thirties because they lived in the forests for days and sometimes weeks at a time. They traveled through them on horseback, packing their supplies with them because there were no roads. These men had a vision of *forests* forever, not the clear-cuts and monocultures that abound today in the name of "forests."

Things changed, however, following World War II, with the invention of the chainsaw and other mechanized ways of leveling forests and converting them into economic crops in the agricultural mode. Today, the old ideals of forestry, as I heard them told while a boy, have given way to the money chase. And my grief continues.

With this in mind, it's now time to give you a sense of today's growing forest crisis—a crisis we humans are passing to the next generation, and every generation thereafter, if we elect to continue down the path we, as a society, are now following, a path based solely on the "dollar chase," both at home and abroad. If we choose to continue along our current, economic path, the forest legacy we leave our children and our children's children will be far less promising than the one we inherited.

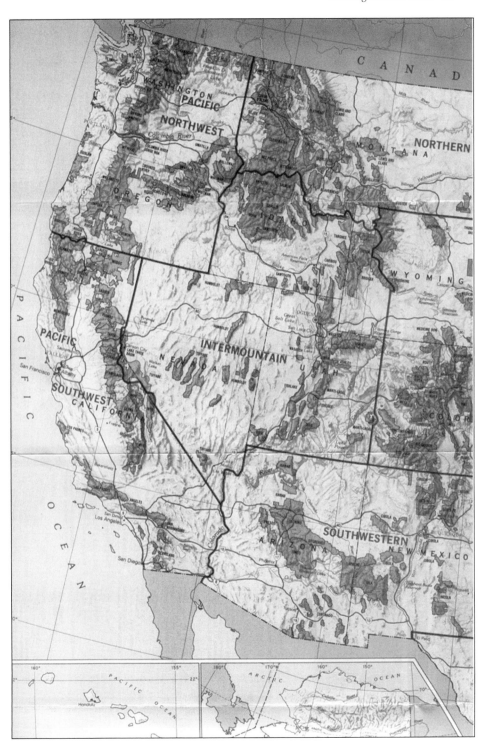

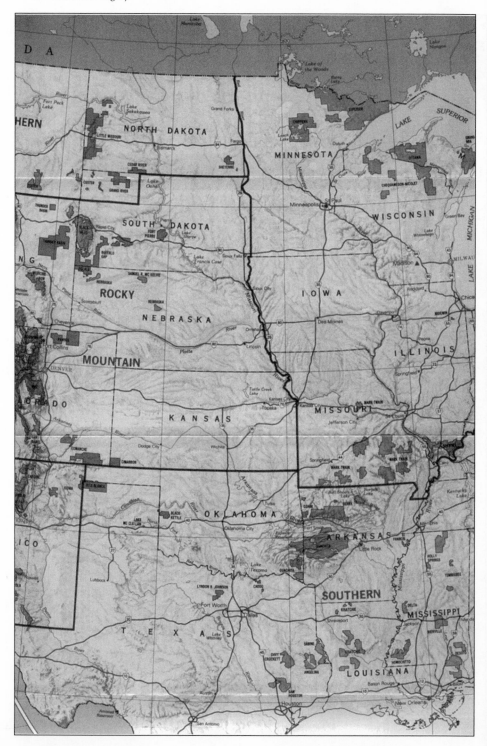

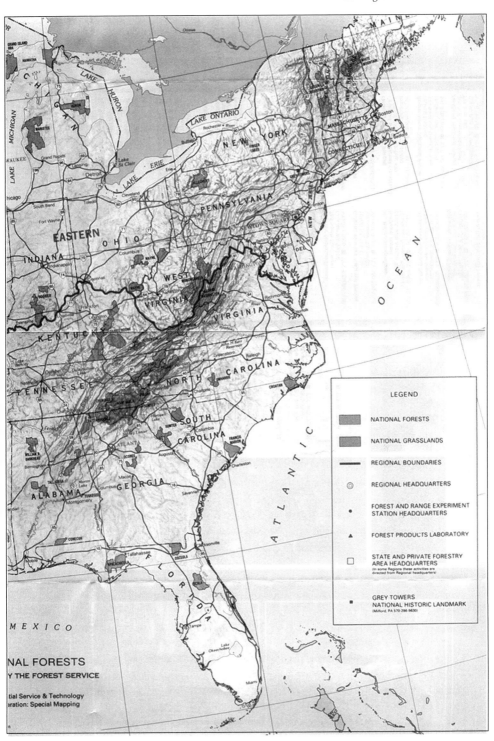

LEGEND

NATIONAL FORESTS

NATIONAL GRASSLANDS

REGIONAL BOUNDARIES

REGIONAL HEADQUARTERS

FOREST AND RANGE EXPERIMENT
STATION HEADQUARTERS

FOREST PRODUCTS LABORATORY

STATE AND PRIVATE FORESTRY
AREA HEADQUARTERS
(In some Regions these activities are
directed from Regional headquarters)

GREY TOWERS
NATIONAL HISTORIC LANDMARK
(Milford, PA 570-296-9630)

NAL FORESTS
Y THE FOREST SERVICE

tial Service & Technology
aration: Special Mapping

Chapter 2
What Drives Today's Forest Crisis?

Call a thing immoral or ugly, soul-destroying or a degradation of man, a peril to the peace of the world or to the well-being of future generations, as long as you have not shown it to be "uneconomic" you have not really questioned its right to exist, grow, and prosper.

— E. F. Schumacher

What drives today's forest crisis? In a word—"economics," *poor* economics, I might add. I say "poor" economics because both "ecology" (which represents Nature) and "economy" (which represents humanity) have the same Greek root *oikos*, a house. Ecology is the knowledge or understanding of the house. Economy is the management of the house. And it is the *same* house, a house we humans have divided at our peril. At issue here is "whether the environment is part of the economy *or* the economy is part of the environment."[12]

Most of the problem with the economic point of view espoused by business people, according to Thomas Gladwin, director of the University of Michigan's corporate-environmental management program, is that business executives and managers often lack good cross-training in science. In fact, less than one percent out of 1.2 million articles written by business professors includes the words "pollution," "air," "water," or "energy."[13]

Because there is such a dearth of good cross-training within science and between science and the liberal arts, the reasoning behind today's forest economics continues to be flawed, which only compounds the complex and confused state of our society with the passing of each decade. The outcome of such flawed economics is that we're so mesmerized by numerical units of labor, production, and commodities that our attention is riveted on manipulating the world for products, all the while we are progressively losing sight of human dignity.

Tragically, it's this neglected area—human dignity—that's the barometer of our well-being. The wholeness, the inner state of self-worthiness of each member of society, is the individual mirror image of society's value system. When that value system faithfully reflects the individual's fulfilled needs for wholeness, then society and its components are in harmony. The paradox is that managing products is a numbers game, whereas human dignity, which is based on spirituality, intuition, creativity, imagination, dreams, and experiences, is not quantifiable.

In a world of rapidly growing population and finite resources, competition becomes severe for whatever is perceived to be limited in supply, the pace of life quickens and yet never seems to be fast enough, all of which leads to frustration that often manifests itself in cynicism. A cynic is a faultfinding critic who plays the role of a helpless victim, believing that human conduct is motivated solely by self-interest and therefore has little or no faith in people. Cynicism is a charge that is often leveled at selected individuals or groups. Inherent in the choice of whether or not to be cynical is the question of having the courage to accept responsibility for one's own actions. A cynic, by definition, projects blame to others for that which is perceived to be missing in oneself.

Another tendency of human beings faced with frustration is to defend a point of view thought to be synonymous with one's survival. There are, however, as many points of view as there are people, and everyone is indeed right from his or her vantage. Hence, no resolution is possible when each person is committed only to winning agreement with his or her position. The alternative is to recognize that "right" *vs.* "wrong" is a judgment about human values that is not winnable in argument. So it's best to define the fundamental issue and focus on it. An issue, usually perceived as a crisis, becomes a question to be answered, and in struggling toward the answer, both positive and negative options not only become apparent but also become a choice.

To the Chinese, crisis means "dangerous opportunity." A crisis, as the word is used in Western industrialized nations, means a decisive moment whose outcome will make a clear difference. Every crisis has two options—positive and negative. For us in the U.S., our crisis is too often in our point of view because we tend to perceive the world through a disaster mentality, regardless of evidence to the contrary.

In field and forest, in desert and ocean, Nature makes the choices. In human affairs, the choice is ours. That we tend to avoid change, to hug our comfort zone, does not diminish the positive option, it only means that we've chosen to argue for our limitations instead of our potential. To opt for potential always involves an element of risk. And we, individually and collectively, tend to focus on and cling to the presumed safety of the status quo, rather than face the emotional discomfort of risking an uncertain, albeit potentially better, outcome.

Although individualism is good, even necessary, in the embryonic stages of an endeavor, it must blend into teamwork in times of environmental crisis. Setting aside egos and accepting points of view as negotiable differences while striving for the common good over the long term is necessary for teamwork. Unyielding individualism represents a narrowness of thinking that prevents cooperation, possibility thinking, and the resolution of issues. Teamwork demands the utmost personal discipline of a true democracy, which is the common denominator for lasting success in any social endeavor.

Even if we exercise personal discipline in dealing with current environmental problems, most of us have become so far removed from the land sustaining us that we no longer appreciate it as the embodiment of continuous processes. Attention is focused instead on a chosen product, the success or outcome of management efforts. Anything diverted to a different product is considered a failure.

It's a critical time to re-evaluate our philosophical foundation of forestry, society, and how we integrate the two. And it's time to re-emphasize human dignity in our decisions. With a renewed focus on human dignity—present and future—as a "product" of the resource decision-making process, we can broaden the philosophical underpinnings of caring for the ecological sustainability of forests and grasslands, oceans and societies rather than only the exploitation of a few selected commodities they produce. Emphasis on human dignity will help foster understanding and teamwork that in turn nurtures mutual trust and respect rather than the "us against them" syndrome.

The "us against them" syndrome exists because our dreams are too small; they're limited only to our own interests and therefore appear separate and in conflict. For example, I want old-growth trees, *or* I want wood fiber, *or* I want wilderness, *or* I want spotted owls, *or* I want native trout, *or* I want clean water, *or*, *or*. . . . What we need instead is a collective dream large enough to encompass and transcend all our small, individual dreams in a way that gives them meaning and unity.

If we dare to dream boldly enough, our special interests will both create and nurture the whole—a healed, healthy, sustainable forest that includes old-growth trees, *and* wood fiber, *and* wilderness, *and* spotted owls, *and* native trout, *and* clean water, *and*, *and*. . . . Such an ecologically sustainable forest is one in which the biological divestments, investments, and reinvestments are balanced in a way that the forest can be self-maintaining. To this end, our public forests are the perfect classroom and laboratory in which to learn and practice caretaking, thereby elevating our notion of forests and forestry in general.

Now, more than ever, we must recognize that we are part of the human family and trust and respect one another as if human dignity was truly the philosophical cornerstone of our American society, as opposed to mere poetic words on an enshrined, national document in a museum. We must also recognize and accept that ultimately we have one ecosystem—often called the biosphere, the "living sphere"—that simultaneously produces a multitude of products. And we, as individuals and generations, as societies and nations, are both inseparable products of and tenants of that living system, custodial trustees for those who follow.

Today I see even more clearly that the past, the present, and the future are all now, in this instant contained, and that the choice of what we do, of where we go is ours. And if perchance we make a mistake, we can always choose to choose again. As long as there is choice, there is hope. As long as there is hope, the human spirit will ever aspire to its highest achievement. It seems appropriate that I now revisit the forest, because I see it not only as the oldest living entity on Earth but also as a potential pillar of our human survival.

This book, *Our Forest Legacy*, is intended as a gift of ideas, and because a gift is free and without conditions, I have no expectations of you the reader. I know that I cannot convince you or anyone else of anything, so I'm not even going to try. I say this because for me to convince you that I am right, I must simultaneously convince you that you're wrong, at which point I have summarily stolen your dignity, and all you can do is defend yourself. If you find something of value between the covers of this book, you may extract it and use it as yours; if not, that too is okay.

Regardless of how you feel, there comes a point in the history and evolution of every individual, profession, agency, and society when change is necessary if that individual, profession, agency, or society is to continue to evolve. And it all begins or ends with the willingness of the individuals—who collectively are the profession, the agency, and society—to change.

Historian Arnold Toynbee asked the critical question: Why did 26 great civilizations fall? The answer, he concluded, is that the people would not, or believed they could not, change their way of thinking to meet the changing conditions of their environment. Thus, with a rigidity of thinking and overpopulation, they stripped the productive capacity of their land and forests, conforming always to the currently acceptable knowledge of the day. At such a juncture stands today's profession of forestry, which can move forward only to the extent that individuals within the profession accept new philosophies and practices demanded by a rapidly changing culture.

A nation without the means of reforming itself is a nation without the means of survival. The same can be said of a profession—any profession. The profession of forestry cannot remain the same and survive, so those who cannot or will not accept new ideas are destined to fall by the wayside.

Physicist Albert Einstein faced a similar situation. He said that he felt as though the ground had disappeared from under his feet when he was faced with quantum physics and the uncertainty principle, because it meant that the "modern perception" of reality was "wrong." All that he had been taught to believe was "wrong." The whole basis of physics had to be rethought in relation to a new set of principles, which didn't guarantee that the new set of principles were correct, only different from those that proved to be incorrect.

If it's difficult for students of basic physics to rethink their discipline, it's even more so for foresters, because their concern must increasingly go beyond the ecological aspects of the forest to include social aspects as well. A forester's reality, which must include and account for the relationship of people to Nature and Nature to people, must once again be readjusted because both knowledge and social values have changed dramatically over the last decade.

Society and science are today challenged to devise a new theoretical framework based on a new set of metaphysical assumptions from which we can restructure our social institutions—our worldview. Pioneers in forestry glimpsed this need for rethinking our social foundation during the reign of the conservation ethic. They sensed that humanity's role in Nature was in error. But much of this understanding was lost in the struggle to build a scientific basis for forestry, because the historical, metaphysical assumptions about Nature were inconsistent with the perception of Nature held by modern science. Ours, therefore, is a spiritual crisis, a perceptual crisis, in which we've lost our sense of place in and of Nature.

To understand how our linear-minded, American practice of economic reasoning has both created and perpetuates today's forest crisis, we must examine, albeit briefly, five things: (1) ramifications of the soil-rent theory, (2) the difference in meaning between "sustained" and "sustainable," (3) the chronic over-exploitation of resources, (4) the linearity of our thinking, and (5) the forest crisis as a product of our thinking.

Ramifications of the Soil-Rent Theory

Today's exploitive forestry practices are based on the precepts of plantation management as the sum total of the discipline we call "forestry," a practice founded on the "soil-rent theory," which calculates the species of tree with the highest rate of internal monetary return and the shortest financial rotation on a given site. Said differently, by holding everything in Nature as a constant value, except the age at which the trees are to be harvested (the age at rotation), one can calculate that age of harvest, or "rotation" in forestry parlance, that will give the highest rate of return on the economic capital invested. The soil-rent theory—a classic, liberal, economic theory—is a planning tool devised by German economist Johann Christian Hundeshagen in the early 19th century for use in maximizing profits as a general objective of economic activities (economically sustained yield).[14] Since its adoption by early foresters, it has become the overriding objective for exploitive "forestry" worldwide.

Forest "management," the dream and illusion of being in "control" of a "regulated" forest designed to operate within "predictable economic parameters," has been, and still is, based on at least six flawed assumptions stemming from the soil-rent theory:

First, is the notion of "absolute freedom." The concept of absolute freedom is inapplicable to forestry (or anything else, for that matter) because all things are defined by their relationships to everything else, relationships that are constantly changing, which makes it physically *impossible* for an "independent variable," such as a tree, to exist in an interdependent world. Ergo, all things—including trees—are constrained within Nature's myriad ecological influences to which they are exposed and to which they must respond. A forest, therefore, is defined by how it functions as an interactive whole, not by the isolation of an economically and/or politically desirable piece, such as a particular "crop" of trees in a particular acreage of soil on a particular company's timberland.

When dealing with scale (e.g., a large, landscape-level forest as opposed to ten acres of planted trees), scientists have traditionally analyzed large, interactive systems in the same way that they have studied small, orderly systems, mainly because their methods of study have proven so successful. The prevailing wisdom has been that the behavior of a large, complicated system could be predicted by studying its elements separately and by analyzing its microscopic mechanisms individually—the traditional linear reductionist-mechanical thinking predominant in Western society (including agriculturally oriented foresters), which views the world and all it contains through a lens of intellectual isolation. During the last few decades, however, it has become increasingly clear that many complicated systems, like forests, do not yield to such traditional analysis.

Instead, large, complicated, interactive systems seem to evolve naturally to a critical state in which even a minor event starts a chain reaction that can affect any number of elements in the system and can lead to a "catastrophe." Although such systems produce more minor events than catastrophic ones, chain reactions of all sizes are an integral part of the dynamics of a system. According to the theory called "self-organized criticality," the mechanism that leads to minor events is the same mechanism that leads to major events. Further, such systems never reach a state of equilibrium, but rather evolve from one semi-stable state to another, which is precisely why sustainability is a moving target, not a fixed end point.

Not understanding this, however, analysts have typically blamed some rare set of circumstances—some exception to the rule—or some powerful combination of mechanisms when catastrophe strikes, again often viewed as an exception to the rule, such as an epidemic of defoliating insects that "suddenly" appear in a plantation of trees.

Be that as it may, a system as large, complicated, and dynamic as a forest can break down under the force of a mighty blow as well as at the drop of a pin. Large, interactive systems perpetually organize themselves to a critical state in which a minor event can start a chain reaction that leads to a catastrophe, after which the system will begin organizing toward the next critical state.[15]

Another way of viewing this phenomenon is to ask a question: If change is a universal constant in which nothing is static, what is a natural state? In answering this question, it becomes apparent that the balance of Nature in the classical sense (disturb Nature and Nature will return to its former state after the disturbance is removed) does not hold. For example, the pattern of vegetation on the Earth's surface is usually perceived to be stable, particularly over the short interval of a lifetime, but in reality, the landscape and its vegetation exist in a perpetual state of dynamic balance—disequilibrium—with the forces that sculpted them. When these forces create novel events that are sufficiently rapid and large in scale, we perceive them as disturbances, such as a massive, lightning-caused fire during a prolonged drought.

Perhaps the most outstanding evidence that an ecosystem is subject to constant change and disruption rather than in a static balance comes from studies of naturally occurring external factors that dislocate ecosystems, and climate appears to be foremost among these factors. For instance, laboratory studies have suggested the rise in carbon dioxide, a cause of global warming, will stimulate plants, including trees, to grow more abundantly. In these studies, carbon dioxide was viewed as an independent variable. In contrast, a new study, conducted on 128 plots in a California grassland, emulated the array of conditions expected to occur as a result of global warming and found a different outcome.

It has long been known that plants use carbon dioxide to produce food, knowledge that seems complete in and of itself, but in reality is only partial. When carbon dioxide is allowed free interplay in the environment, it can be either good or bad for plants. As it turns out, determining whether carbon dioxide is good or bad depends on what other factors are influencing a plant at any given time. Carbon dioxide by itself can increase the growth of plants, but when influenced by companion factors of changes in weather during climatic cycles, such as some combinations of precipitation and temperature, it can actually reduce the stimulus for plants to grow.[16] Moreover, the factors influencing weather and climate are complicated and unpredictable almost beyond belief.[17]

The effect of such free interplay of environmental influences makes it *impossible* for an "independent variable" to exist in an interdependent world because "independent" means "absolute freedom," which itself cannot exist. And because absolute freedom cannot exist, neither can a "constant value," be it economically derived (such as a sustained yield of timber) or otherwise. To be a constant value it would, by definition, have to be *totally* independent of, absolute free from, everything else; again, a physical impossibility.

Second, linear-minded foresters who see only the economic conversion potential of trees into products, such as lumber, assume that, once the indigenous forests are liquidated, they can be replaced by planted "forests" and/or "fiber farms" that are forever renewable on a continual, short (less than 100-year) plant-cut-plant-cut cycle. "Managing" a forest in this manner further assumes that trees constitute a variable independent from the ecosystem itself—provided enough herbicides, pesticides, and fertilizers are applied. The problem with this assumption lies in the linearity of the forester's thinking, as well as the linearity of their predictive models, when married to a cyclic forest that occupies a sphere of unpredictable novelty vastly different in time, space, and function from the linear, economic models of today's exploitive forestry.

Third, agriculturally oriented foresters fail to realize or refuse to accept that their "management" is primarily directed toward what they see above ground, but they cannot alter the ecosystem above ground without simultaneously altering it below ground, a fact largely omitted from the management equation.[18] In turn, they fail to understand or refuse to accept that each tree, each stand of trees, each forest is a mirror reflection of the soil's ability to grow that particular tree, stand, or forest *just one time!*

Fourth, forest productivity rests on five, obvious ecological factors: (1) depth and fertility of the soil wherein the forest grows, (2) quality and quantity of the precipitation reaching the forest, (3) quality of the air infusing the forest, (4) amount of solar energy meeting the forest, and (5) stability of the climate. Foresters committed to the strictly utilitarian, exploitive model of forestry make the mistake in their economic assumptions and planning models, based on the soil-rent theory, of assuming these factors to be constant, despite the extant data on the variability of climate change and its associated fluxes in weather patterns and precipitation. Even if they try to include these ecological variables within their predictive, mathematical models, they do so in a linear mode because it is the only way we humans know how to build such a model. We simply cannot deal with the myriad unknown and incalculable variables that constitute an ecosystem in the present or predictions of how it might function in the future.

Ignoring the unknowability of ecological variables, foresters with an linear, economic mind-set act as though soil, water, air, sunlight, and climate are constant values, which allow a planted crop of trees to be manipulated as an independent, economic variable—a contrived entity that exists only as a conceptual illusion. Regardless of how a forester might think, these five factors remain variables that are never completely predictable.

Soil, for example, is eroded in two ways, chemically and physically; we humans are doing both. We pollute the water and the air with chemicals.[19] In fact, modern, large-scale atmospheric pollution may well have begun with the diffusion of lead from the medieval metal industry a thousand years ago in northern Europe, well before the Industrial Revolution.[20]

Air pollution directly affects the forest by altering the quality of its soil and water, as well as the quality and quantity of the sunlight that drives the forest processes.[21] The chemicals we dump into the air also alter the climate and so the environment that sustains the forest. The interdependent quality and quantity of soil, water, air, and sunlight interacting with a forest are all variables that, in turn, are dependent on the time-scaled cycles of the climate.

The *fifth* error is the conscious simplification of the only part of the forest we humans can readily see, that which is above ground. Linear-minded foresters view their aboveground manipulations as a short-term, economic expedient toward their primary objective of maximizing the production of wood fiber. In pursuit of their objective, they not only eliminate as much biological diversity as possible, such as "undesirable" species of plants and animals (and, unknowingly, their ecological functions), but also eliminate as much "undesirable" genetic diversity as possible.

The *sixth* error is clinging to the capitalistic idea that, for an economic endeavor to be healthy, it must be everexpanding. We, in Western industrialized society, attack the world's renewable, natural resources from an ecologically exploitive, "sustained-yield" perspective that produces increasingly finite limits on most, if not all, potentially renewable resources. So the question becomes one of whether an economically *sustained* yield is—or ever was—possible as a variable independent from the long-term ecological health and integrity of an ecosystem as an interactive whole.[22] If this question is examined from a historical perspective, the answer is a resounding "No!"

The Difference in Meaning Between "Sustained" and "Sustainable"

To understand the difference between "sustained" and "sustainable," it is necessary to consider public lands across a broad scale of time because, while you and I can *physically return* to a given place at our discretion, *we can never go back in time*. That is why the scale of time makes the care and protection of public lands so critical; it represents intergenerational continuity.

We personally get any moment in time precisely once in our lives, and the Universe gets that same moment precisely once in eternity, which makes it absolutely necessary to differentiate between the meaning of "sustained" and "sustainable." Put in terms of forest economics, "sustained" ostensibly refers to an unending yield of timber to be harvested at a given rate as a "non-declining, even flow," a concept the timber industry has pushed over the years. In reality, "non-declining, even flow" is short-term, economically driven, and fixed in time. Moreover, it connotes a *sustained level of cut* based on the *existing* "inventory" of commercial-aged, standing trees (volume of wood)—even though that level of cut is not "biologically sustainable" over the long term when the forest as a whole is taken into account. The discrepancy in the use of and interpretation of "sustained" versus "sustainable" is the fundamental flaw with the "Multiple Use Sustained Yield Act of 1960."[23]

Although the "Multiple Use Sustained Yield Act" may have been of good intent, it is, nevertheless, a misuse of language based on an economic assumption that is totally at odds with ecological reality. The assumption is that ecological processes in a forest ecosystem remain constant, even as we humans strive to maximize whatever product or amenity of the forest seems immediately desirable. Be that as it may, errors in the practice of central-European forestry over the past couple of centuries illustrate well the results of ignoring ecological realities while attempting to maximize short-term profits, based on the "certainty" of economic assumptions. Coupled with the false certainty of economic assumptions is the failure to recognize that a forest is ultimately controlled by inviolate ecological principles, rather than by human desires, scientific knowledge, or technological ingenuity.

There has been some recognition of this failure in Germany, based on historical errors in the practice of forestry. Richard Plochmann, a professor at the University of Munich and District Chief of the Bavarian Forest Service, commenting on German Forestry in 1989, indicated that the concept of forestry would henceforth include mixtures of two or three species of trees, as opposed to the historic, single-species monocultures. In such mixes, at least one species must be indigenous. The ages of rotation (the age at which trees are cut) will depend on their highest value: quality wood as opposed to the historic, inferior, fast-grown wood. This will require rotations of 120 to 140 years. Wherever possible, natural regeneration, as opposed to genetically selected or manipulated seedlings from nurseries, will be used; therefore, clear-cutting will be replaced by shelter-woods in which some of the mature trees are left to reseed the area, cutting trees in small groups, or selectively harvesting trees. Further, herbicides will no longer by used, and insecticides and fertilizers will be used only rarely. Finally, there will be no highly mechanized operations within the plantations as they are brought closer and closer to the physical structure and biological functions of a real forest.[24]

No such changes have been made in the United States, however, and the failure to recognize the sanctity of Nature's inviolate ecological principles, such as the Second Law of Thermodynamics (which states that the amount of energy in forms available to do useful work can only diminish over time), points to the wisdom of changing our thinking and our actions in how we treat forests for the long haul. Forests, after all, are not the indefatigable producers of commodities and amenities that we have heretofore assumed them to be. Instead, forests are living systems that require constant nurturance if they are to produce the products and amenities we want, just as we must nurture our bodies as living systems if we are to remain healthy.

Before we can change our European, utilitarian paradigm, in which we view forests and all they contain simply as commodities to be endlessly exploited, we must sever ourselves from the old, comfortable way of thinking in order to devise a new paradigm. In this new paradigm, we must view a forest as a living organism with which we *cooperate* and so are *allowed* to harvest products as the ecological capability of the forest permits. But first, it is necessary to understand the history of resource over-exploitation.

Chronic Over-exploitation

According to a past popular song, *Me and Bobby McGee*, "freedom's just another word for nothing left to lose." The song, in a peculiar way, points out that we humans have become slaves to our possessions. When unconscious of something's material value, we are free of its psychological grip. Yet, the instant we perceive its material value, we also perceive the psychological pain of its potential loss, a fear that fosters the over-exploitation of our inherited natural resources, such as forests, because not to exploit them is perceived as losing the opportunity, as well as the coveted material value, to someone else. And it's this feeling of loss that we humans fight so hard to avoid. In this sense, it is more appropriate to think of resources "managing" us as opposed to *us* managing resources.

Here we are confronted by three questions if our public forests are to be ecologically sustainable within current economic theory: (1) Is our economic construct, which emphasizes continual growth, possible? (2) Is our economic construct, which emphasizes continual growth, desirable? (3) Will the relentless, acquisitive nature of our economic construct work harmoniously with respect to the ecological sustainability of our public forests?

Along with these questions, we must understand that scarcity can occur in two ways: running out of something (e.g., oil) and overusing something (e.g., our nation's forests). Either way, the current rate at which a given resource is used in today's world, such as our public forests, threatens to exceed the carrying capacity of that resource if the current level of consumption continues unabated. In a world threatened with growing scarcity due to an increasing overpopulation of humans, the general principle of first concern must be the sustainable productivity of the threatened resources, such as our public forests. To counter this notion, linear-minded foresters, and the timber companies they work for, have long justified their linear thinking with the adage: "We plant ten trees for every one we cut," as though the number of trees can somehow replace the relationship of trees to one another in time and space as they constitute a forest.[25]

Historically, then, any newly identified resource is inevitably overexploited, often to the point of collapse or extinction. Its over-exploitation is based, first, on the perceived rights or entitlement of the discoverer-exploiters to get their share before someone else does and, second, on the right or entitlement of the exploiter to protect a perceived economic investment. Such perceived "entitlement" was succinctly expressed by 38-year-old Daniel Hughes, an unemployed logger who steals trees from the Olympic National Forest in Washington. As he sees it: "There are a lot of trees out there, [so] it's easy to get away with this. To me it's like, 'This land is your land and this land is my land.' I'm taking my share." With respect to protecting old-growth trees on public lands for the endangered northern spotted owl, Hughes said: "Yeah, we saw one [once]. We tried to kill it."[26]

Hughes is representative of chronically unemployed loggers seething with resentment over restrictive measures used to conserve the viability of public forests for future generations. These loggers generally feel they are entitled to what they steal. In essence, the world has changed around them, but they *refuse* to change.

While the perceived rights or entitlement of the discoverer-exploiter seem simple enough, there is more to it than this because the concept of a healthy capitalistic system is one that is ever growing, ever expanding, but such a system is not biologically (or even economically) sustainable. With renewable natural resources, such non-sustainable exploitation has a "ratchet effect," where ratchet means to constantly, albeit unevenly, increase the rate of exploitation of a resource to maximize profits as a hedge against the uncertainty of life.

The ratchet effect works as follows: During periods of relative economic stability, the rate of harvest of a given renewable resource, say timber, tends to stabilize at a level that economic theory predicts can be sustained within the market through some scale of time. Such harvest is almost always excessive because narrowly focused, commodity-oriented economists take existing unknown and unpredictable ecological variables, such as fertility of the soil and stability of the climate, and convert them, in theory at least, into known and predictable ("constant") economic values of the projected market in order to better calculate the expected return on a given investment from a sustained yield. To make their theories work, these strictly commodity-oriented economists must "discount" the value of those same resources to and for future generations.

Discounting means translating the values of future social-environmental impacts into equivalent values in today's dollars. While this in fact is what "discounting" means, when it is done on a purely product basis as though all resources belong solely to this generation, it ignores the importance of long-term ecological processes that are subsequently left out of the valuation, such as the unseen contributions that a particular stand (an artificially delineated group) of trees might make to the health of a water-catchment and thus a community's supply of potable water. In effect, such discounting nullifies the future in all economic calculations in order to justify current over-exploitation. In this way, we discount the value of the resources we want today and thereby rob future generations of resources that will have even higher value when that future time becomes the present moment.

Then comes a sequence of good years in the market, or in the availability of the resource, or both, and additional capital investments are encouraged in harvesting and processing because unfettered, competitive economic growth is the root of capitalism. When the inevitable happens and conditions return to normal or even below normal, industry, having over-invested in processing facilities, appeals to the government to increase the "permissible" harvest despite in-place, ecological constraints, because substantial economic capital is at stake—inevitably tied to a *threatened* loss of jobs if the increased harvest is denied.

If the government responds with direct subsidies (monetary) or indirect subsidies (changing the rules by interpreting existing laws in favor of the industry in question), it encourages continual over-exploitation because the industry can continue "business as usual" without fear of losing money. Subsidies paid for non-sustainable exploitation of natural resources bears out the observation English poet and philosopher Samuel Taylor Coleridge made in 1830: "In politics, what begins in fear usually ends in folly." Should the government do nothing to help the industry, the latter often tries to keep its current level of profit by passing the losses to the consumer, which includes their employees.

Suppose, for the sake of discussion, that a company produces hydroelectric power to sell. In years past, when water was plentiful and electricity was relatively cheap, the company made money by encouraging the use of ever-greater volumes of electricity, which increased everyone's use of and dependence on the availability of hydroelectric power. Now, with water scarce because of a prolonged drought, this same company has been encouraging its customers to conserve electricity by using less, the consequence of which has been a reduction in the volume of electricity sold. Then, because the company is not selling as much electricity, the people in charge want to increase the price to maintain their old profit, but in so doing they penalize the very people who made the conservation program work in the first place—the consumers, which includes their employees.

The ratchet effect is thus caused by unrestrained economic investment that increases short-term yields in good times and the subsequent opposition to losing those yields in bad times. In effect, there is opposition to using current resources sustainably for two reasons: (1) because someone else might get the "surplus," which is seen as a personal economic loss and (2) having over-invested when business was booming, there is fear of losing yields and thus profits. This fear of potential loss translates into a great resistance to using a resource in a biologically sustainable manner because there is no visible predictability in yields and no guarantee of yield increases in the foreseeable future. In addition, our linear economic models of ever-increasing yield are built on the assumption that we can in fact have an indefinite, economically sustained yield, even when confronted with ecological constraints that point unerringly to the contrary. The assumption aside, this contrived concept *always* fails in the face of reality, the long-term, biological sustain*ability* of that yield.

Then, because there is no mechanism in our linear economic models of ever-increasing yield that allows for the uncertainties of ecological cycles and their variability, or for the inevitable decreases in yield during economically bad times, the long-term outcome is either a heavily subsidized industry (e.g., agriculture and industrial forestry) and/or a collapsed resource base. The effect is the same, whether in forests or oceans.

Consider this scenario: Over the past half century alone, the catch of cod, tuna, haddock, flounder, hake, and other ocean-dwelling species has dropped more than fifty percent, despite a tripling in the efforts to catch them. This means that over-exploitation of fish in the North Atlantic has been severe enough to undermine Nature's ability to sustain further fishing, according to Dr. Daniel Pauly of the University of British Columbia's Fisheries Center.

A similar circumstance exists along the Pacific Coast of the United States, where a panel of state and federal regulators, known as the Pacific Fisheries Management Council, considered closing the entire continental shelf to commercial trawling in 2003 because of chronic over-fishing. Trawlers catch fish by dragging nets across the ocean bottom, a type of fishing that is widely suspected of causing a drop of forty percent in the allowable catch of commercially targeted bottom fish from 1997 to 2003.

Additionally, data analyzed in 2003 from five ocean basins revealed that the world's oceans had lost over ninety percent of such large predatory fish as tuna, blue marlins, and swordfish, among others, since the advent of industrialized fishing. Clearly, the emerging situation is dire and depicts potentially severe consequences for the ecosystem.[27]

Just the same, when the "specter" of sustainability arises in the context of potential lost profits, parties tend to marshal all scientific data favorable to their respective sides as "good" science and discount all unfavorable data as "bad" science. Such polarization is the seedbed of destructive environmental conflict that, in turn, becomes the stage on which science is politicized, largely obfuscating its service to society.

Because the availability of choices dictates the amount of control we feel we have with respect to our sense of security, based on the sustainability of our profits, a potential loss of money is the breeding ground for environmental injustice. This is the kind of injustice in which the present generations of adults steal from all future generations by overexploiting a resource to maintain and/or increase their profits rather than face change and the subsequent uncertainty of giving up potential income.

There are important lessons in all of this for whomever takes care of our public lands. *First*, history suggests that a biologically sustainable use of any resource— such as soils[28] or forests[29]—has never been achieved without first over-exploiting it, even in the face of historical warnings and contemporary data. If this reading of history is correct, resource problems are not environmental problems but rather human ones that we have created many times, in many places, under a variety of social, political, and economic systems.

Second, the fundamental issues involving resources, the environment, and people are complex, value laden, and process driven, such as our gathering human effects on the world's forests. As humanity has evolved, increased in population, migrated around the world, and progressively altered forests with concomitant requirements for products and services, forestlands have been subjected to continual human-caused change. Virtually no untouched forests remain on our planet, yet people yearn for the ideal of a pristine Garden of Eden, a natural Shangri-La, and so attempt to create it, most often through preservation.

That notwithstanding, as human disturbances invade the remaining forests, they become increasingly fragmented with the accompanying inability to regenerate and perpetuate themselves, provide adequate, quality habitat within and among species of plants and animals. They also continue to be converted to simplified plantation-type "forests" that lack most of the ecological characteristics of a real forest. Moreover, agriculture is still replacing forests, as are urban and industrial developments. With the exception of urban and industrial development, most other kinds of human alterations can be reversed to some extent, but never fully because of compounding, external influences, such as the ever-increasing array of pollutants being dumped into the environment—most notably the air, where they travel unfettered throughout the world and negatively affect every ecosystem with which they come into contact.

An integrated knowledge of multiple disciplines, both scientific *and* social, is required to understand such complex interactions. These underlying complexities of the ecological systems, including social and economic components, preclude a simplistic approach to either resource allocation or the resolution of conflicts among parties competing for the resources. In addition, the wide natural variability and the compounding, cumulative influence of continual human activity tends to mask the results of over-exploitation until they are severe and increasingly irreversible.

Third, as long as the uncertainty generated by continual change—and so the need for new choices with uncertain outcomes—is considered a condition to be avoided, nothing will be resolved. Conversely, once the uncertainty of change is accepted as an inevitable, open-ended, creative life process, most decision-making is simply common sense, and common sense dictates that one would favor actions having the greatest potential for reversibility, as opposed to those with little or none. Such reversibility can be ascertained by monitoring results and modifying actions and policies accordingly—not by simply throwing money at the perceived problem, as governments are wont to do.

Fourth, sustainability is a continual process of choice based on human values that are either aligned with Nature's ecological principles or counter to them: the former based on process-oriented thinking and the latter based on product-oriented thinking. Having said this, I must point out that the answer to one of our most pressing questions still evades us: Will the ecosystems of the future, which we are shaping today, continue to function in such a way as to produce the quality of human life we have come to expect?[30]

Fifth, the seed of all destructive conflict is a perceived loss of choice over our own individual destinies, a loss we interpret as a threat to our survival and to our sense of values. Is the forest legacy we bequeath the next generation and those beyond to be founded on this fear of loss?

In any case, while our knowledge of the way Nature works has increased dramatically since our dawning as a species, we shall never fully know all the answers, even though we will continue to pursue the kinds of information that hint at them. In the meantime, the world continues to change in ways we do not expect and cannot predict, and we change our ideas, albeit slowly, as we experience the ever-changing present. The single factor that seems to make change so excruciatingly difficult to accept for us in the United States is the predominant linearity of our Western industrialized thinking.

The Linearity of Our Thinking

If our democratic society is to last through the twenty-first century, we must begin now to balance the opposing inner stresses: the linearity of our product-oriented intellect and the intuition of our process-oriented heart. I'm pointing this out because I find that people think predominantly in one of two ways.

Some think in a linear pattern that causes them to define ever-increasing production and the accumulation of material products as "progress," hence acquisition as the primacy of life. Such thinking culminates in the degradation of both public and private lands through destructive competition for economic gain. Others come more from the heart and think in a cyclic pattern that causes them to focus on being an integral part of the processes that for them constitute a spiritual participation with life's cycles, a state of consciousness that brought about the creation of our public lands in the first place, as well as their protection through time.

Either way, a society's culture is the expression of its dominant mode of thinking, the means whereby people choose to unite in harmony with the land or progressively exploit it. Given the same piece of land (say, a forest), each culture would produce a different design on the landscape as a result of the dominant pattern of its thinking, a pattern that is the template of personal values expressed in the collective social mirror—the environment:

> Mind is the Master power that moulds and makes,
> And Man is Mind, and evermore he takes
> The tool of Thought, and, shaping what he wills,
> Brings forth a thousand joys, a thousand ills:—
> He thinks in secret, and it comes to pass:
> Environment is but his looking-glass.
>
> — James Allen[31]

Social values determine a culture, and the culture is an expression of those values, wherein it's not the intellect that matters most, but rather the personal character that guides it. And so it is not surprising that the care taken of land by a people is a mirror image of the dominant forces in their social psyche. These secret thoughts, guarded either in the intellect of the mind or the "knowing" of the heart, ultimately express themselves and determine how long a particular society survives or, conversely, how soon it becomes a closed chapter in the history books.

Today, we may well be standing at this crossroads with respect to our forests, and our environment in general, because we, in the United States, treat the Earth as a commodity bank to be endlessly exploited due to the predominant linearity of our collective thinking. Linearity is the doctrine of progress represented by the European invaders crossing oceans and continents, annexing everything in their path for material gain. They used the land, abused it, and moved on, for there seemed always another hill over which to go. Now we, in the present, with no more hills on Earth over which to go, are moving into space on a course of no return, a course that is supposed to bring us into a human-made, material paradise unfettered by Nature's ecological laws.

Our linear-minded society moves through time, discarding old experiences as each new one is encountered. The upshot is that we seldom learn from history because, in our minds, we *never* "repeat" the old mistakes. In reality, we repeat them constantly. We, nevertheless, deny our repeated mistakes in our blind drive for material progress because there is but one definition of progress in our current economic vision—an endless galactic journey of discovery, conquest, and exploitation. To return is to come back to the used and the discarded, while progress is to forever exploit the new.

Within the purview of this endless galactic journey, anything that does not directly contribute to a desired end-product is considered irrelevant, even a waste of resources, time, and energy. The results of such thinking is a disregard for many of the cultural and humanistic values that contribute to the quality of life, an example of which is the massive funding of the scientific portion of our culture (particularly in our universities and state colleges), while the liberal arts, the humanistic portion of these same universities and colleges, languishes.

Too often, in my experience as a government scientist, have I seen the results of this imbalance. It appears in a person having technical skills but few or no interpersonal skills. It is also evident in the exceedingly poor use of the English language, both in speaking and in writing. In fact, when I was in graduate school, I tried to get permission to take a "creative-writing" course for graduate credit. While my major professor agreed that it would be a good thing, I was told by someone higher up in the graduate school that such a course was not worthy of either graduate credit or me as a graduate student.

A characteristic of such linearity is the notion that anything is justifiable as long as and insofar as it is immediately and obviously good for something else— its "conversion potential." What, we ask, is it good for? What can it be converted into? And only if it proves to be immediately good for something other than what it already is, are we ready to raise the question of its value: How much money is it worth? We ask because, since it can only be good for something else, it obviously can only be worth something else, e.g., a mink has value only when converted into a fur coat, a tree has value only when converted into boards.

An excellent example of such linear-mindedness, in the environmental sense, is the valuation of species based on strict economics. Because linear vision looks fixedly straight ahead with the premise that for an economic endeavor to be healthy it must be everexpanding, it views any species that hinders such expansion as necessarily expendable. Further, if a species cannot be converted into something else, it has no value: take the case of the Pacific yew, a small shrub-like tree thought to be a "weed" and treated as such before it became the only source of *taxol*, a drug used to treat breast cancer in women. Once the chemical essence of taxol was deciphered and synthetically produced, the Pacific yew was again relegated to being nothing more than a "weed."

Linearity flourishes in contempt for the processes on which it depends. In doing so, it blinds us to the forest by allowing us to see only the trees, not the biological processes that produce and maintain the forest—of which the trees are but a visible expression. Linearity is founded on the simplistic, artificial destiny prescribed by economic formulae pertaining to those things that can be converted into money, but has not a whit to do with the irreplaceable, irretrievable novelty of living organisms, let alone species or whole ecological systems.

Processes are invariably cyclic, rising and falling, giving and taking, living and dying in space on ever-expanding ripples of time. Yet linearity places its emphasis only on the rising phase of the cycle—on intellectual knowledge, production, expansion, possession, youth, and life. It relentlessly shuns intuition, return, idleness, contraction, giving, old age, and death.

In contrast, cyclic vision recognizes our life as an endless repetition of basic and necessary patterns in the circular dance of use and renewal, expansion and contraction, life and death. This vision sees everything as interdependent, as fulfilling the ecological excellence and uniqueness of its function, which means, as I've already said, there is no such thing in Nature as an "independent variable, " a "constant value," or "absolute freedom." Everything in the Universe is patterned by its interdependence on everything else, and it is the dynamic pattern of interdependence that produces the novelty of change—the universal constant with which we must interact, whereby we grow, and through which Nature survives.

Whether we see change as friend or foe depends on how we think about it intellectually and how we feel about it intuitively. Change is the great crucible in which we must learn to balance our intellect and our intuition, our linear drive for material products and our nurturance of the cyclic processes that produce the products in the first place.

Herein lies our greatest challenge for the twenty-first century with respect to our environment: to balance our drive for linear production with recognizing the need for and honoring the constraints of the cyclical, ecological processes that create and maintain the integrity and health of our collective environment. In the end, we must understand and *accept* that the process *is* the product, and the product *is* the process. One cannot exist without the other, a unity that means personal growth and social evolution cannot exist without change. In this case, we must first put to right our inner landscapes before anything will change in the outer landscape, such as the ecological health of our public forests, something Aldo Leopold touched on when he wrote in 1928:

> The American public for many years has been abusing the wasteful lumberman. A public which lives in wooden houses should be careful about throwing stones at lumbermen, even wasteful ones, until it has learned how its own arbitrary demands as to kinds and qualities of lumber, help cause the waste which it decries.[32]

If our society is to survive the twenty-first century with any sense of human dignity, our social-environmental decisions must be predicated on protecting the integrity of the ecological processes that support our quality of life and that of the generations of children, born and unborn. Here, the imperative is how our thinking affects our nation's forests, indeed the forests of the world.

This being the case, there is need for a new vision, for an elevated level of consciousness in dealing with the continually emerging environmental issues rooted in the old, worn-out thinking of standard, exploitive forestry. These issues are multi-faceted, interrelated, cross all political and bureaucratic boundaries, and are challenging the very survival of humanity itself, as attested in a statement penned by one hundred Nobel Laureates—among them the Dalai Lama, Mikhail Gorbachev, and Archbishop Desmond Tutu—at the 2001 Peace Prize Centennial Symposium in Oslo, Norway. Their statement concludes: "To survive in the world we have transformed, we must learn to think in a new way. As never before, the future of each depends on the good of all."

The Forest Crisis as a Product of Our Thinking

Our linear focus in the face of cyclical processes further exacerbates the way our thinking affects the viability of our public forests. Clearly, how we think determines what we do. Exploitive forestry's narrow, economic focus on wood fiber, in whatever form, is destroying our nation's forest.

A narrow, economic focus on the individual pieces of a system is like a racehorse with blinkers that prevent it from seeing anything but the racetrack in front of it. Such people tend to disregard any piece of the system for which they see no immediate personal value. This linear-minded thinking occurs in isolation of how the system itself really functions. A person oriented to seeing only the immediate effect of desirable pieces of a system usually denies that removing or altering a perceived desirable piece as an individual act, or as an act by an individual human, can or will negatively affect the productive capacity of the system a whole. This is but saying linear-mindedness often creates a problem within a dynamic, interactive, cyclical system by trying to create and control an independent variable—a physical impossibility, as the following example illustrates.

The Dayak people of Borneo had such a problem with malaria in the 1950s that the World Health Organization sprayed huge amounts of DDT over the countryside in order to kill the malaria-carrying mosquitoes. While the mosquitoes were indeed killed, so was the population of tiny, parasitic wasps that kept the population of thatch-eating caterpillars in check. With the caterpillars out of control, the thatch roofs of the Dayak people's houses collapsed. In addition, the DDT poisoned other insects that were eaten by the geckos that, in turn, were eaten by the domestic cats, causing them to die, which allowed the rat population to explode and threatened the Dayak people with outbreaks of typhus and bubonic plague. In response to the demise of the cat population, the World Health Organization engaged a squadron of the British Royal Air Force stationed in Singapore to parachute 14,000 live cats into Borneo, to once again alleviate a symptom—the explosion in the population of rats seen and treated as an independent variable. This whole scenario came about because the malaria-carrying mosquitoes were seen and treated as an independent variable in an *interdependent* system.[33]

In contrast, a systems-thinker (a person who sees the whole in each piece) is concerned about tinkering willy-nilly with the pieces because they know such tinkering might inadvertently upset the desirable function of the entire system. Benjamin Franklin, in his "Poor Richard's Almanac," expressed it simply: "For want of a nail, the shoe was lost. For want of a shoe, the horse was lost. For want of a horse, the rider was lost. For want of a rider, the battle was lost."

Yet, just because one person is a mill owner, economist, industrial forester, or engineer does not necessarily mean that person is a narrow, economic thinker, any more than the fact that another person is an ecologist or an environmentalist necessarily means that person is an outright systems-thinker. Moreover, there is nothing inherently "right" or "wrong," "good" or "bad," "better" or "worse" with respect to linear-mindedness *vs.* systems-thinking. They are simply two architects with differing points of view who, left to their own devices, would each design a different landscape on the same piece of ground, each depicting opposite extremes of a thought continuum. This continuum goes from the most cyclical/process-oriented terminus (*consummate systems-thinker*) to the most linear/product-oriented terminus (*narrow, economic focus*) and contains *all shades in between*.[34]

How we think can be symbolized as: S<————————>P, where "S" equates to the most cyclical/process/systems-oriented terminus of the continuum and "P" equates to the most linear/product/economically oriented terminus. In terms of how one thinks, a systems-thinker is a process-oriented generalist who is good at synthesis, while a linear-minded specialist, who derives intellectual knowledge about discrete pieces, supplies the information required for the synthesis. The specialist endeavors to study, and thereby "understand," the elements of the puzzle in isolation of their interactive function, whereas the generalist attempts to fit the elements together in order to form a functional whole.

The clash between linear-minded, product-oriented thinkers and process-oriented systems-thinkers is a conflict of values based on different world views, something science can approach only indirectly, politicians tend to studiously avoid, and economists most often paint with bold, linear, Napoleonic strokes of certainty in an uncertain world. After all, one's perception is one's truth, and the "truth" one clings to normally makes it difficult to hear, much less accept, anything new. By way of illustration, let's consider two competing scenarios: the exploitive branch of the timber industry with its insistence on cutting the remaining old-growth trees and the environmentalists who work to protect them.

Why the timber industry's dogmatic insistence on cutting every old-growth tree? The answer is simple. The industry has nothing invested in the growing of the old trees, and so every old tree represents a "windfall profit." The industry's only investment is associated with getting the old trees to the mill. In this sense, the timber industry knows exactly what it wants—old-growth trees. How many it wants—all of them. When it wants them—now. Why it wants them—because they're essentially free for the taking. Where it wants them—at the mill. That intact, old-growth forests are, in many ways, critically important to the quality of life for the generations of children, both those here now and those yet to come, is simply not a consideration. The profit margin is all that matters; so the future is discounted. To make this point clear, I offer a short quote from Karl F. Wenger, President of the Society of American Foresters:

> The fact is, Nature knows nothing. Nature is deaf, dumb, blind, and unconscious. . . . It reacts blindly and unconsciously according to the properties and characteristics of its components. These have no intrinsic values, since only the human race can assign values. Nature doesn't care what we do to it. . . . Clearly, the people's needs are satisfied much more abundantly by managed than by unmanaged forests.[35]

There are many environmentalists, however, who see an intrinsic value in the remaining old-growth trees and want to protect them. These folks perceive the old trees to be important to the ecological health of the forest as a whole *and* to the spiritual health of society as a whole—present and future. This sense of valuation goes far beyond the present and, rather than discounting the future, it actually produces an accruing ecological value over time, which amounts to an investment account of systems productivity for the generations yet to come.

There is, however, a little-considered, devastating outcome to this on-going battle. While both sides argue over who gets the product, the old trees, we are losing the forest because the processes that form the embodiment of the forest as an interactive system are being summarily ignored by the narrowness with which the battle lines are drawn. Without caretaking the processes, because they are out of focus in the ensuing battle, they fall into disrepair and the health of the forest declines accordingly. Moreover, the old trees are irreplaceable in the near future should scientific data prove them to be an irrefutable, ecological necessity to the welfare of the forest and thus the welfare of society at large.

These competing values set up an unfortunate dichotomy: A systems-thinker usually *responds* to a changing circumstance and is willing to focus on transcending an issue in whatever way may be necessary to frame a vision that protects the system itself for the good of the majority in all generations. The flip-side of the coin is that a linear-minded thinker normally *reacts* to a changing circumstance and dogmatically fights to protect the status quo, to avoid change at almost any social-environmental cost, particularly when and where money is involved.

Consider, for example, that in May of 2002, residents of John Day, a cattle and timber town in northeastern Oregon, attempted to steal the jurisdiction of the Malheur National Forest from the U.S. Forest Service for private monetary gain. These folks have hard, narrow attitudes toward "outsiders" and "the government," meaning the federal government. By a margin of about two-to-one, residents passed a measure to allow private people to *commandeer* the "right" to cut trees on public land without permission of the U.S. Forest Service—or you and me, the owners of the forest.

The measure states that Grant County residents may "participate in stewardship of natural resources on public lands within the County, when those resources or the use of those resources becomes detrimental to the health, welfare or safety of the people." The backers of the measure maintain that it will give local people license to cut hazardous or fire-prone trees on federal lands in the name of public safety, with or without permission from the U.S. Forest Service.

This measure came about because the vast over-cutting of the national forests, especially during the administration of President Ronald Reagan (1981-1989), left loggers to watch their livelihood trickle away. Still, the people blamed timber policies and environmental restrictions for keeping them off public lands that had given them jobs as loggers, truck drivers, mill workers, and ranchers. In any event, they failed to recognize that the policies and restrictions they objected to were in *response* to the over-cutting *they* were party to. Without these policies and restrictions, there would be no viable forest for future generations.

According to supporters of the measure, millions of board feet of timber could be "salvaged" if only they were allowed to cut the big (old-growth) ponderosa pines and firs that are "hazards." Dave Taylor, who helped draft the measure, was of the opinion that: "If we could just address salvage on the dead, dying, and blowdown, we could provide a lot of trees to the mills."[36]

While this may sound reasonable, I have over many years seen how the linear-mindedness inherent in the notion of "salvaging" "hazardous" trees has resulted in pilfering the largest and most economically valuable trees in a forest one at a time, even when it meant bulldozing a temporary skid trail a quarter of a mile off of a road to reach one large tree. And then there is the issue of salvage itself, especially since it is based on the economic premise that any merchantable tree not taken to the mill is a "waste."

There is, however, no such thing in a forest as "waste" in the biological, systemic sense. The economic concept of waste has spawned the industrial precept of salvage logging, an activity that discounts non-monetary social values and all intrinsic ecological values. The danger of salvage logging lies primarily in its philosophical underpinnings that justify the perception of immediate economic considerations to the exclusion of all else.

Salvage logging disrupts the interdependent relationships of the forest by removing all merchantable dying and dead trees. To illustrate, a live old-growth tree eventually becomes injured and/or sickened with disease and begins to die. How a tree dies determines how it decomposes and reinvests its biological capital (organic material, chemical elements, and functional processes) back into the soil and eventually into another forest.

How a tree dies is important to the health of the forest because its manner of death determines the structural dynamics of its body as habitat. A tree may die standing as a snag to crumble and fall piecemeal to the forest floor over decades. Or, it may fall directly to the forest floor as a whole tree. Regardless of how it dies, the snag and the fallen tree are only altered states of the live tree, as I said before. Consequently, the live old-growth tree must exist before there can be a large snag or a large fallen tree.

Structural dynamics of a dying or dead tree, in turn, determine the biological/chemical diversity hidden within the tree's decomposing body, greatly affecting the ecological processes that incorporate the old tree into the soil from which the next forest must grow. What goes on inside the decomposing body of a dying or dead tree is the hidden biological and functional diversity that is totally ignored by economic valuation. Consequently, the fact that trees become injured and diseased, die, and remain in place is critical to the long-term structural and functional health of a forest.

The forest is an interactive, organic whole defined not by the pieces of its body but rather by the interdependent functional relationships of those pieces — the intrinsic value of each piece and its complementary function as they interact to create the whole. While harvesting live trees based on silvicultural prescriptions is designed to make money, it does so within at least some planned ecological constraints; whereas salvage logging is *reactive* to keep from losing potential monetary gains. As such, salvage logging amounts to the unplanned, opportunistic extraction of trees without ecological constraints. It must be remembered, however, that the lack of prudent, ecological constraints is simply a matter of choice.

I understand the fear of change faced by the people of John Day because I made a career change myself and started over, without prospects, when I was 48 years old. Be that as it may, is it morally justifiable to commandeer for private use a public forest held *in trust* for *all American citizens in all generations*, rather than risk changing careers?

Commandeering the Malheur National Forest for private use points out the connection between the simplistic narrowness of a liner-minded approach to life's uncertainties. The greater a person's propensity for linear-mindedness, the more certain they are of their knowledge and the more black-and-white their thinking tends to be. This intellectual constriction may have led American psychologist William James to observe: "A great many people think they are thinking when they are merely rearranging their prejudices." Clearly, the "John Day Measure" is just one more example of the chronic over-exploitation of our public forests for which all generations will pay the compounding costs—unless we choose to change our thinking.

The level of consciousness that created a problem in the first place is not the level of consciousness that can fix it. We must, therefore, raise our level of consciousness with respect to how we opt to caretake our public forests. We, as a people, as a society, are *not* "locked" into our way of thinking. Because we, as a people, as a society, can always choose to choose again, there is hope for the generations of the future.[37]

Chapter 3
Our Public Forests

Here again it is Time, which does not plead for one side or the other but is the infallible judge in the last resort, which will have the last word.
— Louis Pasteur, Chemist

Whereas short-term economic expediency—privatizing profits and commonizing costs—has been a mainstay of human society, its environmental consequences probably first became evident in the forests of the world. Even President Franklin Delano Roosevelt observed that "We treat timber resources as if they were a mine." In his book, *A Forest Journey*, John Perlin presents a comprehensive history of global deforestation and its consequences.

The book begins in ancient Mesopotamia 4,700 years ago, where civilization first emerged and people started their extensive exploitation of forests, and closes in late 1800s in the United States, the last Western nation to leave the age of wood. By that date, the United States had already lost most of its eastern forest, one of the greatest ever to grow on Earth. Consequently, reliance on wood as the primary fuel and building material gradually gave way to coal and iron.

Perlin ends his book with a quote from an 1882 issue of *Harper's Monthly*; the article was written by N. Egleston, one of America's leading authorities in forestry:

> We are . . . following . . . the course of nations which have gone before us. The nations of Europe and Asia have been as reckless in their destruction of the forests as we have been, and by that recklessness have brought themselves unmeasurable evils, and upon the land itself barrenness and desolation. The face of the earth in many instances had been changed as the result of the destruction of the forests, from a condition of fertility and abundance to that of a desert. . . . The masses of the people . . . should have set before them the warnings of history.[38]

Why Our Public Forests are Important

In 1908, President Theodore Roosevelt convened the first meeting of all the governors of the states to address the topic of the environment. His opening address to the conference is as pertinent today as it was then. He began:

> I welcome you to this Conference at the White House. You have come

hither at my request, so that we may join together to consider the question of the conservation and use of the great fundamental sources of wealth of this Nation.

So vital is this question, that for the first time in our history the chief executive officers of the States separately, and of the States together forming the Nation, have met to consider it. . . .

This conference on the conservation of natural resources is in effect a meeting of the representatives of all the people of the United States called to consider the weightiest problem now before the Nation; and the occasion for the meeting lies in the fact that the natural resources of our country are in danger of exhaustion if we permit the old wasteful methods of exploiting them longer to continue.

Later in his speech he said, "Just let me interject one word as to a particular type of folly of which it ought not to be necessary to speak. We stop wasteful cutting of timber; that of course makes a slight shortage at the moment. To avoid that slight shortage at the moment, there are certain people so foolish that they will incur absolute shortage in the future [in this case, self-centered, linear-minded timber barons], and they are willing to stop all attempts to conserve the forests, because of course by wastefully using them at the moment we can for a year or two provide against any lack of wood." He went on to say that "Any right-thinking father earnestly desires and strives to leave his son [and daughter] both an untarnished name and a reasonable equipment for the struggle of life. So this Nation as a whole should earnestly desire and strive to leave the next generation the national honor unstained and the national resources unexhausted. . . ."[39]

Gifford Pinchot, the first chief of the U.S. Forest Service, also recognized this point when he wrote in his 1914 book *The Training of a Forester*:

> The forest is a national necessity. Without the material, the protection, and the assistance it supplies, no nation can long succeed. Many regions of the old world, such as Palestine, Greece, Northern Africa, and Central India, offer in themselves the most impressive object lessons of the effect upon national prosperity and national character of the neglect of the forest and its consequent destruction.[40]

Public lands are important, if for no other reason than, by inference, they already have been designated as public trusts owned by the American people at large for the present *and* all future generations, a designation resoundingly dismissed by most—but not all—people in the timber industry and the George W. Bush Administration. If the generations of the future are to have viable forests (Nature's classroom) wherein to learn how to maintain healthy forests and to restore those degraded through decades of exploitive practices, forests on public lands must be consciously cared for as a collective and living trust in order to ensure, as much as humanly possible, their ecological sustainability as an unconditional gift from one generation to the next.

Before I can write about caretaking public forests, however, it is necessary to discuss "our forest legacy," the one we leave for the benefit of future generations. Although it has been suggested by an anonymous reviewer that I will waste my time as writer and yours as reader by dissecting the meaning of "our," "forest," and "legacy," I beg to differ. If there is a single element inherent in being human, it is our often-ambiguous use of language (e.g., "sustained" vs. "sustainable") as previously noted, which enflames unnecessary conflict over misunderstood expressions of personal values that directly affect the overall health of our shared environment.

What's in a Word?

When I speak of language, I am referring to the ability to say what we mean and mean what we say because, in the best of all worlds, clarity in understanding is the most precious of gifts to receive and the most difficult to give. While the written word can appear to be a tangible expression of a particular sound (the "word" itself), the meaning it conveys remains an abstraction of human thought. Moreover, it is not incumbent on you, as reader, to understand what I write, but rather it is my duty as conveyor of ideas to be as clear as I humanly can in my writing. Part of that clarity must be an examination of the meaning ascribed to the verbal bricks I use in constructing the foundation of this book because their beauty, utility, and durability are dependent on how precisely they are chosen, configured, fitted together, and held in place by the syntactical mortar of my literary skills.

To this end, I—and you—must understand, as best we can, what is meant by the terms "our," "forest," and "legacy" because those three words are implicit in the heart of caretaking our public forests. That's simple, you may think; just go to the dictionary and look them up. Well, if we look up "our," we find the following definition in the 1999 *Random House Webster's Unabridged Dictionary*: "a form of the possessive case of *we* used as an attributive adjective: *Our team is going to win*. Cf. ours." If we now consider "ours," we find the following: (1) "a form of the possessive case of *we* used as a predicate adjective: *Which house is ours*?" and (2) "that or those belonging to us: *Ours was given second prize*."

If we look up "forest," we find: (1) "a large tract of land covered with trees and underbrush; woodland;" (2) "the trees on such a tract; *to cut down a forest*;" and (3) "a tract of wooded grounds in England formerly belonging to the sovereign and set apart for game." The Society of American Foresters has described a *forest* as "a plant community predominantly of trees and other woody vegetation."[41] And the International Union of Forestry Research Organizations offers three definitions: "1. (ecology) Generally, an ecosystem characterized by a more or less dense and extensive tree cover; 2. (ecology) more particularly, a plant community predominantly of trees and other woody vegetation, growing more or less closely together; and 3. (silviculture/management) An area managed for the production of timber and other forest produce, or maintained under woody vegetation for such indirect benefits as protection of [water] catchment areas or recreation."[42] Although these brief definitions have been offered by foresters and ecologists, each is a thoroughly simplistic characterization of a forest.

What about "legacy?" It is defined in Webster's as: (1) "*Law*, a gift of property, esp. personal property, as money, by will: a bequest" and (2) "anything handed down from the past, as from an ancestor or predecessor: *the legacy of ancient Rome*."

What does this tell us about "our forest legacy?" Not much, because the universe we humans perceive through our senses is indefinable, no matter how hard we try to make it otherwise. Why? Because words like "our," "forest," and "legacy" are only symbols, metaphors for something we perceive that, nevertheless, defies a direct approach either through touch or intellectual discernment. It is therefore impossible to "define" anything through language, in part because there are no words for the language of the heart. For this reason, a forest can be variously characterized, but never defined because a forest—as does everything else—touches the totality of the Universe, most of which is far beyond our meager comprehension. Nevertheless, I once wrote a book, *Forest Primeval: The Natural History of an Ancient Forest*, trying to define what I thought to be two, simple words: "ancient" and "forest."

Taken alone, their meanings were difficult enough to corral, but taken together, the potential transference of the symbology concealed in "ancient forest" became impossible. Accordingly, it should come as no surprise that I not only failed to define either word singly but also together because I could not, and cannot, define "Universe," or "life," or "death," or "time," or "tree," or "soil," or "fire," or "Earth"—each an interactive part of the forest. I could only assign descriptors to selected parts of the concepts I perceive each word to convey. As in the dictionary, I could only address what I thought I understood. On the other hand, I could not, and cannot, define what the above words convey as a whole because I can neither define the interrelatedness of the Creative processes embodied in infinite becoming, nor the processes' infinite novelty in space or time, and most certainly not in space *through* time. Further, the meaning of a given word may change through continual usage and so help to confound intergenerational communication.

In other words, be it the Universe, the Earth, or a tree, all things are ever in motion, always in the infinite fluidity of Creation, never static, never in the finite of the concrete, of the created. In my endeavor to define a forest as an entity, I discovered that the intellect of my mind sees but a fraction of the world and is incomplete in and of itself. Further, each entity has an unseen history and an unforeseeable future, so how can I define something that includes infinite motion in time and space, infinite change, with a finite history but an infinite and unknowable novelty? I cannot, so I must use a symbol or a metaphor to reach beyond the limits of my understanding and the boundaries of language.

Dictionary definitions and scientific definitions can be changed, laws can be promulgated, interpreted, reinterpreted, and rewritten, but spiritual values and those of well-being, once lost, are irreplaceable. And it is this sense of irreplaceability that a forest represents, especially an old forest that has marked many human generations in its march through the centuries and, in some cases, millennia. Such trees are the bristlecone pines in the Great Basin National Park of Nevada, where they have stood experiencing life for over 3,000 years. One, now cut, was over 5,000 years old—one of the oldest living beings on Earth.

While I gave intellectual definitions of a forest in the beginning of this book, those kinds of definitions have blurred for me as I studied forests and learned to know them as living entities with whom I feel a deep, spiritual connection. When my sense of spirituality is added to the intellectual construct of my science, with all it definitions and their nuances, I begin to understand the struggles of a poet when using words in an effort to reach beyond the limitation of words. And so, for me, there are no "concrete" definitions of anything spiritual—only intuition and feelings. That notwithstanding, I shall be as succinct for you as possible in defining what I mean.

There is, however, one aspect of language that defies me completely, no matter how hard I try to overcome it, and that is to share my feelings. I spoke earlier of my love for the forest and of my deep, abiding grief at witnessing its continual, and often mindless, destruction worldwide, but especially in North America, most particularly public forests in the United States, where we have enough scientific data to mandate saving them for their intrinsic values, which ultimately sustain us all. Because the forests are a vital necessity for the whole of humanity through time, they are part of the global commons—in other words, everyone's birthright. What arrogance allows anyone to presume to "own" a forest with the "right" to destroy it, perhaps for all generations? Beyond that, by what arrogance does an administration of our government allow anyone to destroy our public forests, enshrined as they are in legislative trust for all of us as American citizens—present *and* future? And why do we, the American people, allow it to happen? Are "we the people" not "the government?"

Are we too preoccupied with the money chase to grieve? Are we too numb with daily concerns, terrorist attacks, and the perpetual wars to grieve? Are we too overloaded by the media's obsession with violence to grieve? Are we too frightened of criticism to grieve? Are we too unconscious to grieve? I don't know. What I do know is that my grief compels me to speak out and to continue speaking because the only way out of grief is to think and act as psychologically mature adults who genuinely care about one another and the legacy we bequeath those who inherit the world we leave—the children. With this in mind, I'm going to consider the notion of "legacy" as I think it pertains to caretaking our public forests.

Our Forest Legacy

What does "our forest legacy" mean? Granted, we have already defined the words individually, but what do they mean in sequence? What does "our" refer to? Does it refer to "forest," or "legacy," or both?

The possessive "our" refers only marginally to "forest" because we did not, and cannot, create a forest in the sense of being in control of Nature, so we cannot *own* a forest outright. It cannot be *ours* in an unconditional sense—despite the fact that we may possess a legal deed and title to the land. A deed and title notwithstanding, we can take care of a forest as a custodian, guardian, trustee, etc., which means we can manipulate, for better or worse, what Nature has already created, but within the non-negotiable ecological limitations Nature imposes on our ambitions. In this limited, "hands-on" sense, we can think of the forest as *ours*.

But there is another sense of "ours," a shared sense borne of humanity's common need for the products and amenities of the world's forests if humanity is to survive with any sense of well-being and dignity. The required products, such as water and oxygen, are concrete entities that are non-substitutable. On the other hand, spiritual renewal through a connection with the oldest living beings on Earth—the "great grandparent trees"—is considered an amenity by those who see the forest as their "bread basket" while engaged in the money chase, where "amenity" is defined in the dictionary as: "any feature that provides comfort, convenience, or pleasure."

Despite the intellectual definition of amenity, to me, and I'm sure to others, my spiritual connection with the forest is as much an irreplaceable, non-substitutable necessity as is the oxygen and water the forest provides. Therefore, the "ours," in connection with public lands means *every* American citizen, regardless of where they live, where they came from, what they believe, who they are, whether born or unborn. And it is this sense of "ours," that which we cannot afford to lose, that for me underpins the notion of "legacy" in "our forest legacy."

Can we *own* a legacy? Can it be *ours* in the truly possessive sense? Yes, it can, because, in the context of this book, "legacy" (in the sense of "forest legacy") is the sum total of our knowledge of a forest and our thoughts and decisions with respect to that forest as they are manifested in the ecological outcomes of our actions, whatever they are. In turn, it is the ecological outcome of our decisions and subsequent actions that we bequeath the next generation as the circumstances with which that generation must contend.

In this sense, *our* "forest legacy" is embodied in the degree to which a forest is healthy and that, in turn, equates to the level of ecological sustainability we pass into the future, beginning with the next generation—our children. Our legacy is also the reflection cast back to us by the social-environmental mirror, the reflection by which our children and our children's children will remember us as they confront the circumstances they inherited based on the care we took of the forests they must live with, just as we had to live with the condition of the forests handed to us by our parents and grandparents.

To this end, we must learn to characterize a system by its function, a concept that recalls from the past Aldo Leopold's perception that "the first precaution of intelligent tinkering is to save all the pieces."

Intelligent tinkering means paying heed to the "precautionary principle," a concept engendered more than twenty years ago when private landowners in Germany noticed their beloved forests were dying. Appealing to the government for action, they incited an all-out effort to reduce acid deposition, most commonly known as "acid rain," from coal-fired power plants.

Based on the perceived validity of the old adage "an ounce of prevention is worth a pound of cure," the German government translated the adage into law: *Vorsorgeprinzip*, which literally means the "forecaring principle." During the ensuing years, *Vorsorgeprinzip* became enshrined in international law as the "precautionary principle."

The precautionary principle, or thinking through the possible consequences of a potential action before committing the act, instructs us humans to acknowledge our mistakes, admit our ignorance, doubt the certainty of our knowledge, and act with humility in order to honor our place as an inseparable part of Nature and of one another through all generations. Such instruction gives us a way to change our personal and collective thinking and behavior for the greatest good for all generations by acting consciously to minimize the harm we cause in the course of our living since nothing in life is free of consequences.

In creating a forest legacy, therefore, one must remember that forestry on public lands is, first and foremost, about human values other than the derivation of monetary profit, but economic profit is by no means mutually exclusive from other social values. Whereas the focus on private, industrial lands is primarily concentrated on an economically sustained yield, the focus on public lands must be on an ecologically *sustainable* yield if the forest legacy is to serve future generations well. To create such a legacy, we adults must consciously protect the rights of the American people to enjoy all the advantages derivable from our public lands (provided they respect those lands and neither injure nor destroy them), as opposed to the individual rights of private and/or semi-private ownership in whatever political form it might take.

In creating our forest legacy, the scientific questions and knowledge that help us understand our public forests in an ecological sense must be reconnected with the value system that governs how "we the people" treat the forests. We *can* reconnect our scientific endeavors with our social values since we are not "locked into" any given social paradigm. We *can* reconnect our science and our social values by dropping the pretense of "objectivity" in our pursuit of science, and use science to understand and honor the ecological principles that govern Nature, our human place in the scheme of things, and thereby work toward the highest ideals of social-environmental harmony we can imagine and pass forward to the next generation.

If we want to leave a forest legacy of sustainable quality as a bequest to our children and our children's children, we must carefully envision what we choose to do in caring for our public forests and how we choose to do it. Designing our forest legacy is best thought of as though our social values, our choices, and our actions were encapsulated within a glass jar. I use a glass jar as a metaphor because, while we can see into it and arrange what we place therein, it has a particular shape and limited space that can accommodate only so many things of various sizes and shapes. The jar itself represents the limits Nature places on our freedom to manipulate the forest sustainably, a freedom that is substantively limited by circumstances and constrained by time. Precisely because our freedom is limited, we must pay particular attention to how we arrange whatever objects we choose to place in the jar since their relationship will determine how they function therein.

Let's pretend we have a collection of variously sized pebbles and a pile of sand that we want to place in a particular glass jar for enjoyment and storage, but their combined volume exceeds that of the jar. How would we arrange the pebbles and sand to get the most in? Keep in mind that there is a greater volume of sand than all the pebbles combined, and there are more small pebbles than large ones, even though the large ones represent more volume than all the small ones.

If we put the sand in first, there will not be enough room for all the pebbles. If we put the sand and some small pebbles in, there will be little or no room for the large pebbles. If, on the other hand, we arrange the large pebbles (our social values) in the jar first and hold them in place with smaller pebbles (our choices), we can fill in the remaining spaces with grains of sand (our actions) and thereby get most of our large pebbles in place, as well as a goodly number of smaller ones and a reasonable amount of sand to secure the arrangement so it can be safely passed on to our children as an unconditional gift.

As we are filling our "forest jar" and creating an ecologically sustainable forest legacy on public lands, we are simultaneously creating an ongoing legacy forest that becomes geometrically more important over time as most private forestlands are continually degraded through standard, exploitive forestry practices.

A Legacy Forest

In both the context of this book and in the context of forestry, a "legacy forest" is one in which an imaginary line is drawn around all or a portion of a forest, in this case a legal boundary that separates public portions of a continuous forest from private portions. Within this boundary, the composition of plant species, the physical structures they provide, and the functional processes these structures allow are passed forward, either as an ever-changing biological snapshot in time and/or as the most "natural" progression of Nature's processes we can protect from direct human interference. Both of these scenarios are used as points of reference against which to evaluate changes wrought by successive human alterations of the surrounding landscape across space and time. In essence, a legacy forest can be likened to an enlarged section of the overall forest blueprint in that natural processes can be studied in greater detail than would be possible when viewing the forest as a whole.

I hope it is clear that a "legacy forest" has a different and narrower connotation than does a "forest legacy." Yet, in both cases "our" is fully applicable in the possessive because the legacy, whatever its form, is something we create through our motives, attitudes, thoughts, decisions, and actions, which crystallizes into ecological outcomes in the forest. We then pass those outcomes to the next generation, either as a particular condition, in a particular place, in a specific time, for a specific reason (as in "legacy forest") or as the general state of health of the forested ecosystem (as in "forest legacy"). Either way, "our," in terms of "legacy" is how we archive the values and ideals we espouse—be it a biologically healthy, sustainable, and economically viable forest, or something of lesser social-environmental quality.

In addition, a legacy forest highlights the way we humans see ourselves today—separated from Nature, but in two different ways. Some people still envision humanity's role in life as the master species whose duty it is to harness Nature strictly for the benefit of human society. Others perceive humanity as an unnatural intrusion into the world of Nature, where the very presence of contemporary humanity is somehow defiling. Here the connotation of "natural" is of something apart from human society, a purity without contamination by human activity or artifact. Neither view is accurate in the face of humanity's obligatory relationship with Nature, a relationship that is both reciprocal and self-reinforcing.

What we perceive as "natural" is not an either/or proposition, but rather a degree of naturalness based on our personal feelings of the sacredness with which we participate with Nature. The way we, as a society, participate with Nature is currently one of our most pressing struggles—to balance spirituality and materialism, intellect and intuition, use and conservation. And our public forests are the stage on which this struggle is currently being enacted.

Unfortunately, this struggle often takes place in the courts of the land rather than in our hearts and souls, where such a battle belongs. In our struggle to find a sustainable place within Nature and to participate with Nature in a sustainable manner, we must recognize and accept that for thousands of years human society has been converting Nature's landscape into one of human design. And we shall continue to do so in perpetuity. It's all we can do, simply because we exist and because we occupy and use the whole world, leaving nothing untouched from the blue arc of the heavens to the deepest reaches of the sea.

Even Indigenous North Americans, who were in spiritual harmony with Nature when the Europeans arrived, were in fact altering the landscapes in which they lived. But they did not change them in the same way as did the Europeans, who invaded and colonized the New World and forcefully superimposed their materialistic demands on the spiritually oriented customs of the indigenous peoples.

Most of us who are Americans of European extraction still act as though we are alienated from Nature. We still don't have a clear sense of "naturalness" as a concept joining us with Nature. We still see our idea of "natural" as somehow excluding us as participants in the creative process of designing the landscapes in which we live and work.

Nonetheless, we are a natural part of the landscape in which we live, even if we don't understand it, and what we do in the way of converting Nature's landscape to our cultural landscape is natural—albeit our activities may be wantonly destructive. The "naturalness" of a landscape is not an absolute value but rather a relative one that ranges from no alteration by humans at the most pristine end of the ecological scale, such as an inaccessible mountain top, to any kind of artificial alteration at its most humanized end, such as a shopping mall.

Our living in and altering of the landscape are natural aspects of Nature's ever-evolving creative process by the very fact that we exist and we are, of necessity, active participants in redrafting Nature's design. Ergo, in the sense of our "naturalness," we both belong here and have a right to be here, in addition to which we have a duty to *consciously* participate in the creation of our landscapes. So the question is: How do we consciously participate in the sacred care of the Earth while we help sculpt and texture its landscapes with our cultural designs?

One way is to use a legacy forest as a benchmark to maintain or create a situation wherein the composition, structure, and function of the system approximates our understanding of the most natural end of the continuum, which, as I said before, allows natural processes to be studied in greater detail than would be possible otherwise. And while a legacy forest can exemplify the most natural end of the continuum, a tree plantation or "fiber farm" constitutes the extreme cultural end.

Such a continuum can easily be symbolized as follows: N<—-—>C, where "N" represents the most natural end of the continuum, in this case the legacy forest, and "C" the most cultural end, the exploitive fiber farm. Everything in between, depending on where it falls along the continuum, represents either a greater degree of naturalness or a greater degree of culturalness, but rarely an absolute balance between the two.

From here on, the subject of this book is healing our public forests and maintaining them in good ecological health. As you read, you will discover that I am writing about an ideal. I do so because the ideal is all that is worth striving for. In this sense, I will describe how I think a forest can be cared for as a "living trust." Although I am writing about an ideal, I am not intimating the kind of Utopia described by Sir Thomas More, that imaginary isle of perfection in human relationships, but rather, I am writing about what I think is possible if we are willing to change our thinking and choose a path into the future that reunites the house of ecology and economy.

Chapter 4
A Forest in Trust

Each word I utter must pass through four gates before I say it. At the first gate, the keeper asks, "Is this true?" At the second gate, the keeper asks, "Is it necessary?" At the third gate, the keeper asks, "Is it kind?" At the fourth gate, the keeper asks, "Is this something you want to be remembered for?"
— An ancient Arab proverb

Much has been written through the centuries about Utopia, which is "no place" until we imagine it, and then it always becomes "some place."[43] The Utopias imagined by philosophers are difficult to reach; they can be glimpsed only after strenuous, focused, intellectual and physical effort. But whosoever has succeeded in making the journey to the shores of possibility on the Idyllic Isle brings back tales of a land where people love one another; where work is transformed into labors of love, or "play" as some would call; and where earthly social-environmental problems, entwined as they are almost beyond recognition, are untangled with patience, compassion, and ease. Earth, too, could be like this, so the story goes, if only. . .

The difficulty with utopias is not that they are *imagined perfection*, but rather that they are *imagined cures for imperfection*, and therein lies the problem. Namely, a solution is conjured in an attempt to move *away* from an unwanted negative circumstance rather than moving *toward* a desired positive outcome. Put another way, instead of moving toward the ideal, most solutions attempt to cure an imperfection by moving away from it, an action that is neither physically nor psychologically possible.

To heal and protect our public forests in an ecologically sustainable condition, we must have a destination in the form of a vision toward which to journey. The ideal can then help define an agenda that rests firmly on the bedrock of a shared vision, one that incorporates the collective wisdom, personal courage, and political will needed to inspire true social progress. Despite the usual elusiveness of Utopia, creating the forest legacy envisioned in the pages of this book is within the realm of human attainability, should people choose to make it so. It is, after all, only a choice and the will to carry it out.

Here I ask you to remember that success or failure is a crisis of the will and the imagination, not of the possibilities. To me, the only real failure is not to risk trying, for clearly, without risk there can be no gain. In fact, success or failure is *not the event itself*, but rather the *interpretation* of the event, as illustrated by the story of Flambeaux:

> Flambeaux left Cut Off, Louisiana, and moved to De Berry, Texas, where he bought a donkey from an old farmer for $100. The farmer agreed to deliver the donkey on the following day.
>
> The next day, the farmer drove up and said, "I'm sorry, but I have some bad news . . . the donkey died last night."
>
> "Well den, Sir," said Flambeaux, "jus' give me money back."
>
> "I can't do that Sir," replied the farmer, "I spent it already."
>
> "OK, den. Jus' unload dat donkey."
>
> "What are you gonna do with him?" the farmer wanted to know.
>
> "I'm gonna raffle 'im off."
>
> "You can't raffle a dead donkey, you dumb Cajun!"
>
> "Well dats where you wrong. You wait an' you learn jus' how smart we Cajuns is!"
>
> The farmer saw Flambeaux a month later and asked, "What happened with that dead donkey?"
>
> "I raffled dat donkey off. I sold 500 tickets at two dollars apiece and made $998."
>
> "Didn't anyone complain?" asked the farmer in disbelief.
>
> "Jus' dat guy who won. So, I gave 'im 'is two dollars back," said Flambeaux with a grin.[44]

A Biological Living Trust

Although most people speak of forest "stewardship," I personally prefer the concept of a "living trust" because "stewardship" does not in and of itself have a legally recognized "beneficiary"—someone who directly benefits from the proceeds of one's decisions, actions, and the outcomes they produce. Although a "steward," by definition, is someone who "manages" another's property or financial affairs and thereby acts as an agent in the other's stead, there is nothing explicit in the definition about a legal beneficiary. For this reason, "stewardship" is a much more wishy-washy term than "living trust" because the fiduciary responsibility of "stewardship" is to the shareholders; whereas the fiduciary responsibility of a "living trust" is to the beneficiaries, none of whom need to be physical shareholders.

A living trust is like a promise, something made today, but about tomorrow. In making a promise we relinquish a little personal freedom with the bond of our word. In keeping our promise, we forfeit a little more freedom in that we are limited in our behavior, but to break a promise is to lose some of our integrity and a bit of our soul. The reason people hesitate to make promises lies in the uncertainty of circumstances on the morrow. Helping to quell the fear of uncertainty is the purpose of a "living trust."

A "living trust," in the legal sense, is a present transfer of property, including legal title, into trust, whether real property (such as forestland) or personal property (such as livestock, jewelry, or interest in a business). The person who creates the trust (such as the owner of forestland) can watch it in operation, determine whether it fully satisfies his or her expectations, and, if not, revoke or amend it.

A living trust also allows for the delegation of administering the trust to a professional "trustee," such as a "managing" forester, which is desirable for those who wish to divest themselves of managerial responsibilities. The person or persons who ultimately receive the yield of the trust, for better or worse, are the legal beneficiaries. The viability of the living trust is the legacy passed from one generation to the next, which means we must think in terms of "potential productivity" *instead* of constant production.

Though a trustee may receive management expenses from the trust, meaning that a trustee may take what is necessary from the interest, at times even a small stipend, the basic income from the trust, as well as the principle, must be used for the good of beneficiaries. In our capitalist system, however, natural resources are *assumed* to be income or revenue, rather than capital.[45] That said, a trustee is obligated to seek ways and means to enhance the capital of the trust—not to diminish it. Like an apple tree, one can enjoy the fruit thereof, but not destroy the tree. A living trust, after all, is about the quality of life offered to the generations of the future; it is *not* about the acquisition of possessions.[46]

Because a forest is a living entity, it can be thought of as a *"perpetual,* biological living trust,"* (hereinafter referred to as "biological living trust") in which individual people—as well as their relationships among one another, Nature, their communities, and generations—have value and are valued, as are all living beings. For forestry to survive throughout the twenty-first century as an honorable profession, it must accept the moral essence of a biological living trust. It must also advance beyond resisting change as a condition to be avoided (clinging to the current, linear, reductionistic, mechanical world view of exploitive forestry) and embrace change as a process filled with hidden, viable ecological-social-economic opportunities in the present for the present *and* the future—the beneficiaries. People with the necessary courage to unconditionally accept change are rare, but I remember meeting one in 1992 in Slovakia.

I had been asked to examine a forest in eastern Slovakia and give the people my counsel on how to restore its ecological integrity after years of abusive exploitation by the Communists. During the process, I worked with employees of the Slovakian Federal Forest Service. One man, the Chief Forester, then near the end of his career, had been in charge of the forest during the days of the Communists. As I was about to leave Slovakia, the Chief Forester took me aside and said, with great emotion: "Chris, if I learned one thing from you, it is that the forest is sacred—not the plan. Thank you." With that, this man reversed the thinking of his entire 40-year career. I have seldom encountered such courage, humility, and dignity.

We all need such courage, humility, and dignity if we are to be worthy trustees of our public forests as a biological living trust. But before trusteeship and the precept of a biological living trust can be fruitfully discussed as a means of caretaking forestlands, you and I must be able to understand and integrate two perspectives of time, that of a clock and that of an hourglass.

Time as measured by the ticking of a clock is constant in tempo. With a clock, you see the hands move from second to second, minute to minute, and hour to hour—as 'round and 'round the clock's face they go. While to a child time seems to drag, even stand still, to an older person time seems to fly, despite the fact that watching a clock's hands make their appointed rounds belies both the impatience of youth and the sensation of time as fleeting in old age.

Contrariwise, if you measure time through the functioning of an hourglass, you have the distinct impression that time is "running out," like the sand pouring to the beck and call of gravity from the top of the hourglass, through the small hole in its middle, to the bottom. Most adults view time with a growing sense that theirs is running out, so they must grab all of life they can before their time is "spent," a fear of loss that champions material acquisitiveness in the supposed "safety" of the status quo. This sense of impending loss as time "runs out" causes people to avoid, as best they can, the inevitable admission of change.

In reality, of course, time does not run out; our bodies expire instead. And it's precisely the dual sense of time running out and the demise of our bodies that causes many people to seek a way to continue their sense of being in the world, like the continual ticking of the clock. One way to accomplish such continuance is through a living trust.

If we have the courage and the willingness to adopt and implement the concept of a "biological living trust," we are practicing sustainable forestry in which ever-adjusting relationships—ecological, social, and economic—become the creative energy that guides a vibrant, adaptable, ever-renewing forestry profession through the present toward the future. After all, forestry could be a profession that constantly opens the mind with growing conscious awareness because the forests of tomorrow will be created out of the inspirations, discernment, choices, decisions, and activities of today. In addition, sustainable forestry honors the integrity of both society (intellectually, spiritually, materially) and its environment, thereby fitting the concept of a biological living trust in that it *maintains* positive outcomes for both the forest as a dynamic system and the beneficiaries who depend on the forest for their well-being.

A biological living trust is predicated on systemic "holism" in which reality consists of an organic and unified whole that is greater than the simple sum of its parts. That is to say, the desired function of a system defines its necessary composition. The composition, in turn, defines the structure that allows the functional processes to continue along their designated courses. Consequently, wisdom dictates that we must learn to characterize a system by its function, *not its parts*. The basic assumptions underpinning a biological living trust—all externalities within the current economic framework of forestry—are:

- Everything, including humans and nonhumans, is an interactive, interdependent part of a systemic whole.
- Although parts within a living system differ in structure, their functions within the system are complementary and benefit the system as a whole.
- The whole is greater than the sum of its parts because how a system functions is a measure of its ecological integrity and biological sustainability in space through time.
- The ecological integrity and biological sustainability of the system are the necessary measures of its economic health and stability.
- The biological integrity of processes has primacy over the economic valuation of components.
- The integrity of the environment and its biological processes have primacy over human desires when such desires would destroy the system's integrity (= productivity) for future generations.
- Nature determines the necessary limitations of human endeavors.
- New concepts must be tailored specifically to meet current challenges because old problems cannot be solved in today's world with old thinking.
- The disenfranchised, as well as future generations, have rights that must be accounted for in present decisions, actions, and potential outcomes.
- Nonmonetary relationships have value.

In a biological living trust, the behavior of a system depends on how individual parts interact as functional components of the whole, not on what each part, perceived in isolation, is doing. The whole, in turn, can only be understood *through* the relationship/interaction of its parts. Hence, to understand a system as a functional whole, we need to understand how it fits into the larger system of which it is a part. This understanding gives us a view of systems supporting systems supporting systems, *ad infinitum*. Consequently, we move from the primacy of the parts to the primacy of the whole, from insistence on absolute knowledge as truth to relatively coherent interpretations of constantly changing knowledge, and from an isolated personal self to self in community. At this point, you might wonder if a forest can be a living trust in the legal sense? The answer is, "Yes."

A Forest as a Biological Living Trust

As he retired from his position as Deputy Chief for the National Forest System, Jim Furnish wrote an open letter to the Chief of the U.S. Forest Service, dated October 2001. The letter opens and concludes with these paragraphs:

It has been my privilege to serve my entire career with the Forest Service, most recently as Deputy Chief for the National Forest System. I would like to close with some personal thoughts to you, my Chief, and to the agency as a whole. I think we stand at an important crossroads in our history. The choices made in the years ahead will have important impacts on our effectiveness and will shape the role that national forests and grasslands play in America and the world. . . .

From my first day in the Forest Service, I have looked at the multiple-use message with the words Wood, Water, Forage, Wildlife and Recreation circling the shield. This simple emblem spoke well to the values of our society and a Forest Service bristling with confidence, striving to meet their needs. It served us well. But we need a new message.

We need to squarely address the issue of sustainability in a way that exhibits humility and an advocacy for the natural resources entrusted to our care. Societies throughout history have come to acknowledge— more often than not too late—that if one takes care of the earth, the earth will provide.

I hunger to see the Forest Service enthusiastically and openly embrace a new set of values for our second century of service. Being clear about what we stood for was what made the Forest Service great. We can emerge once again as international leaders in conservation, but not by being silent.

I share these thoughts with you in hopes that they will help you prosper in your role as Chief. The Forest Service occupies a privileged role in America. The incredible foresight to establish and grow the National Forest System has reaped untold benefits for this great land and people. We are now struggling to assert the legitimacy of our vision as the rightful and proper leaders to manage this estate. We are moving in a new direction, but lack clarity and purpose. I think it is the highest calling of leadership to be clear and explicit about the destination, as well as the urgency to get there.

All we have to offer our children and the generations of the future—*ever*—are choices to be made and things of value from which to choose. Those choices and things of value, both biological and social (= legal), can be held within the forest as a living trust, of which we, the adults of the current generation, are the legal caretakers or trustees for the next generation. Although the concept of a trustee or trusteeship seems fairly simple, the concept of a trust is more complex because it embodies more than one connotation; consider a forest as a legal living trust.

The forest is a biological living trust in the present for all generations. A living trust represents a dynamic process, whether in the sense of a legal document or a living entity. Human beings inherited the original living trust—planet Earth—long before legal documents were invented. The Earth as a living organism is the ultimate biological living trust of which we are the trustees and for which we are all responsible. Our trusteeship, in turn, is colored, for better or worse, by the values our parents, peers, and teachers instilled in us, our experiences in life, and the ever-accruing knowledge of how the Earth functions as an ecosystem.

Even so, the administration of our responsibility for the Earth as a living trust has throughout history been progressively delegated to professional trustees in the form of elected or appointed officials when and where the land has been, and is, held in legal trust for the public—"public lands." In so doing, we empower elected or appointed officials with our trust, our firm reliance, belief, or faith in the integrity, ability, and character of the person who is being empowered.

On public lands, such empowerment carries with it certain ethical mandates that in themselves are the seeds of the trust in all of its senses—legal, living, and personal:

- "We the people," present and future, are the beneficiaries; whereas the elected or appointed officials and their hired workers are the trustees.
- We have entrusted these people to follow both the letter *and spirit* of the law in its highest possible sense.
- We have entrusted the care of public lands (those owned by all of us), whether forested or otherwise, to officials and professionals—planners, foresters, and other people with a variety of expertise, all of whom are sworn to accept and uphold their responsibilities and to act as professional trustees in our behalf.
- Our public lands—and all they contain, present and future—are "the asset" of the biological living trust.
- We, the American people, have entrusted these officials and professionals with our public lands as "present transfers" in the legal sense, meaning we have the right to revoke or amend the trust (the empowerment) if the trustees do not fulfill their mandate: Public lands are to remain healthy and capable of benefiting all generations.
- To revoke or amend the empowerment of our delegated trustees if they do not fulfill their mandates is both our legal right and our moral obligation as individual, hereditary trustees of the Earth, a trusteeship from which we cannot divorce ourselves.
- As U.S. citizens, we have additional responsibilities to critique the professional trusteeship of our public lands because we are taxed to support not only the delegated trustees but also to provide public services with respect to those lands, and elected officials make the dollar allocations on our behalf. Their decisions about where and how to spend "our" money are reflected in both the present and future condition of our public lands.

How might this work if we are both beneficiaries of the past and trustees for the future? To answer this question, we must first assume that the administering agency is both functional and responsible. The ultimate mandate for the trustees, be they employees of an agency or otherwise, would then be to pass forward as many of the existing options (the capital of the trust) as possible.

These options would be forwarded to the next planning and implementation team (in which each individual is a beneficiary who becomes a trustee) to protect and pass forward in turn to yet the next planning and implementation team (the beneficiaries that become the trustees), etc. In this way, the maximum array of biologically and culturally sustainable options could be passed forward in perpetuity.

Should the officials and/or professionals fail to fulfill their obligations as trustees to our satisfaction, their behavior can be critiqued through the judicial system, assuming the judicial system is functional. In this way, the carefully considered effects embodied in our decisions as trustees of today could create a brighter vision for the generations to come. In order for this to happen, however, the notion of a biological living trust must become a "big idea."

A Biological Living Trust as a Big Idea

Real learning—the remembrance of things forgotten and the development of things new—occurs in a continuous cycle. Learning encompasses theoretical and practical conceptualization, action, and reflection, including equally the realms of intellect, intuition, and imagination. Real learning is important because overemphasis on action, one part of which is competition, simply reinforces our fixation on short-term, quantifiable results. Our overemphasis on action precludes the required discipline of reflection, a persistent practice of deeper learning that often produces measurable consequences over long periods of time.

Many of today's problems result from yesterday's solutions, and many of today's solutions are destined to become tomorrow's problems. This simply means that our quick-fix social trance blinds us because we insist on little ideas that promote fast results, regardless of what happens to the system itself. What society really needs are "big fixes" in the form of systemic ideas that promote and safeguard social-environmental sustainability, e.g., a collective vision of our public forests as a biological living trust.

Where, asked the late publisher Robert Rodale, are the "big ideas," those that change the world? They probably lie unrecognized in everyday life since our culture lacks sufficient freedom for general thought.

A "big idea," according to Rodale, must:

- be generally useful in good ways—a biological living trust translates into a healthy environment and available resources;
- appeal to generalists and give them a leadership advantage over specialists—a biological living trust requires an understanding of the system as a whole *and* so necessitates an amalgamation of generalists and specialists, with generalists in charge;
- exist in both an abstract and a practical sense—a biological living trust, as seen in number one above, is practical in its outcome, but it is also abstract in that its practical outcome requires people to work together with love, respect, humility, wonder, and intuition, as well as their intellect;
- be of some interest at all levels of human concern—a biological living trust requires the continual building of relationships, which is all we humans really do in life and so touches all levels of society, both within itself and with Nature;

- be geographically and culturally viable over extensive areas—a biological living trust is a general necessity if the natural world is to remain viable and habitable for the generations of the future;
- encompass a multitude of academic disciplines—to caretake public lands as a biological living trust requires the integration of all disciplines, such as soil science, mycology, philosophy, sociology, theology, education, politics, ecology, forestry, and economics; and
- have a life over an extended period of time—a biological living trust is, by definition, an instrument of continuity among generations.[47]

A biological living trust seems to fit all of Rodale's requirements. It also helps people to understand that life is not condensable, that any model is an operational simplification, a working hypothesis that is always ready for and in need of improvement. When we accept that there are neither shortcuts nor concrete facts, we will see how communication functions as a connective tool through which we can and must share experience, invention, cooperation, and coordination.

When people speak from and listen with their hearts, they unite and produce tremendous power to invent new realities and bring them into being through collective actions. While today's environmental users with narrow, special interests will not be around by the end of this century, all of the environmental necessities will be, and that makes "trusteeship" critically important.

"Trusteeship," in terms of public lands, is a process of building the capacity of people to work collectively in addressing the common interests of all generations within the context of sustainability—biologically, culturally, and economically. A biological living trust, in turn, means honoring the productive capacity of an ecosystem within the limitations of its ecological principles.[48] This said, every public forest can be on a trajectory toward sustainability if we begin now to caretake each of them as a biological living trust, which *is* a "big idea." After all, sustainability is only a choice—our choice, but one that must be carefully and humbly planned if it is to endure the often short-sighted, contradicting political vagaries of humanity.

Remember, to protect the best of what we have in the present for the present *and* the future, we must all continually change our thinking and our behavior to some extent. Society's saving grace is that we all have a choice. Accordingly, whatever needs to be done *can* be—if enough people want it to be done and decide to do it.

Chapter 5
Overseeing the Trust

"Courage" comes from the French word for "heart" or "spirit."

A person who serves the people by working as an employee of the government of the United States, whether as a federal judge, politician, scientist, or caretaker of public forests, must pass the tests described in the eulogy that Senator William Pitt Fessenden of Maine delivered on the death of Senator Foot of Vermont in 1866:

> When, Mr. President, a man becomes a member of this body he cannot even dream of the ordeal to which he cannot fail to be exposed;
>
> of how much courage he must possess to resist the temptations which daily beset him;
>
> of that sensitive shrinking from undeserved censure which he must learn to control;
>
> of the ever-recurring contest between a natural desire for public approbation and a sense of public duty;
>
> of the load of injustice he must be content to bear, even from those who should be his friends; the imputations of his motives; the sneers and sarcasms of ignorance and malice; all the manifold injuries which partisan or private malignity, disappointed of its objects, may shower upon his unprotected head.
>
> All this, Mr. President, if he would retain his integrity, he must learn to bear unmoved, and walk steadily onward in the path of duty, sustained only by the reflection that time may do him justice, or if not, that after all his individual hopes and aspirations, and even his name among men, should be of little account to him when weighed in the balance against the welfare of a people of whose destiny he is a constituted guardian and defender.[49]

Two years after Senator Fessenden delivered this eulogy, his vote to acquit Andrew Johnson brought about the fulfillment of his own prophecy. But then, liberty is always expensive. True liberty demands that each and every leader know when his or her duty to history has been completed and when to step down with dignity and grace so the pivotal idea of a free democracy as the central pillar of our nation can deepen in the centuries to come. It is this kind of democracy that is needed in overseeing our public forests for all generations. Part of this democracy means that each generation must be the conscious keeper of the generation to come—not its judge.

As the guardians of the next generation, it is incumbent on us, the adults, to prepare the way for those who must follow. This will entail, among other things, wise and prudent planning in the caretaking of forests on public lands as a biological living trust.

To this end, the task of those who caretake the public forests is to apply our ecological, social, and economic knowledge, imagination, and systemic-thinking to the fundamental redefinition and redesign of forestry as it is currently institutionalized and practiced to bring it into accord with Nature's ecological principles as we understand them. The redefinition and redesign of forestry—as "whole-system" forestry—is necessary in order to create a cultural foundation in the practice of forestry that both honors and protects social-environmental sustainability. It is, however, concurrently necessary to understand how today's agencies function in order to gain some idea of how to redefine and redesign an agency, such as the U.S. Forest Service, in a way that prevents old, chronic problems from simply reoccurring under a different guise in the near future.

Stages In The Cycle Of An Agency

Although every president has some sort of conceptual vision of what America should be, they inevitably run up against old, established agencies that are loath to change. While working within the Bureau of Land Management, I survived three presidential administrations, each of which attempted to redesign the agency. In fact, President Nixon even tried to combine the U.S. Forest Service with the U.S. Bureau of Land Management, and lost. For someone who has never worked within a bureaucratic agency, it may be difficult to understand how an agency can develop into an entity that refuses to change or die, as well as what stages it goes through to reach that point of seeming "immortality."

I am not aware of any reference that describes the developmental stages of an agency, as has been done for an individual and a family, and I have not taken part in the formation of an agency. I can, however, think of four generalized stages of development that I have personally experienced while working on large projects within an agency. Although the projects were undoubtedly simpler than the creation of an agency itself, some were incredibly exciting, and it is from participating in one of these projects that I can imagine what it must be like to take part in the creation of an agency like the U.S. Forest Service. To this end, I will briefly share my experience of helping to shepherd a large project from its inception through its completion, because I perceive a similarity between the life of a large project and that of a public agency.

Stage one: The inception of an agency is based on a perceived need that is in the public interest. This perception revolves around one person or a small nucleus of people with a vision, as clearly stated by Gifford Pinchot in his book *Breaking New Ground*. He saw the "conservation policy" he helped to forge as the guiding principle of the U.S. Forest Service, of which he was the first chief from 1898 through 1910:

> The Conservation policy . . . has three great purposes.
> First: wisely to use, protect, preserve, and renew the natural resources of the earth.
> Second: to control the use of the natural resources and their products in the common interest, and to secure their distribution to the people at fair and reasonable charges for goods and services.
> Third: to see to it that the rights of the people to govern themselves shall not be controlled by great monopolies through their power over natural resources.[50]

In a letter written on March 5, 1905, Pinchot concisely stated his mission, the guiding philosophy on which the early U.S. Forest Service was founded: "the greatest good for the greatest number in the long run." Yet even with the clearly perceived need for a guardian of the public interest, the formation of the Forest Service was no easy task. There were many bitter, political battles to be fought with men who wanted all the land put in private ownership for their personal gain.

Although the seminal ideals that guide the inception of an agency may be clearly defined for their time and place in history, such as those of the U.S. Forest Service, we now look back and wonder exactly what was meant as we struggle to meet today's perceived "needs" from our national forests. What exactly did Pinchot mean by "the greatest good for the greatest number in the long run?" While one may not agree with his perceived motives, as attested by the political actions taken within the last two decades to increasingly privatize our public lands, it must be noted that whatever the ideal on which the Forest Service was founded and whatever actions were taken to implement that ideal, it was new and daring in its time, and it was meant to be a service held in trust for all the people, both present *and* future.

It is easy to consider the vision of the pioneers of the past as simplistic in today's context because of a greater biological knowledge about forests, a different perception of desires and necessities from forests, a different perception of society, and a vastly greater number of people all vying for what the forest has to offer. While their vision may seem simplistic, it is important to remember those pioneers did the best they could with the knowledge available and the dreams they had in a less complex, historic time. It is also important to keep in mind that today we are the pioneers of the future. Are we doing the best we can with what we know? Will the agencies in which we now serve or that we now create fare any better than those of the past?

Stage two: After its inception, an agency goes through a period of false starts and misfires in searching for an identity. Out of this fumbling can come growth and a joining together to fulfill the vision, provided the vision is clearly stated and firmly agreed to in the first place:

> Pinchot was a great and electric leader by any standard. Stewart Udall called the Forest Service's Washington headquarters during Pinchot's regime "the most exciting place in town." And it was, as Pinchot and Teddy Roosevelt successfully conspired against private western ranching, timber, and water interests to set aside 148 million acres, three-fourths of today's system, as national forests.[51]

One can only imagine what it must have been like in Pinchot's day. I was privileged to experience this type of electric excitement first-hand during the years that I worked in La Grande, Oregon, at the Forest Service research laboratory. We were putting together a book entitled *Wildlife Habitats in Managed Forests: The Blue Mountains of Oregon and Washington* (referred to simply as the "Blue Mountain Book") and were frequently "in over our heads." Part of the time we spun our mental wheels as we struggled to say what we thought we meant and at the same time be sure that we really meant what we thought we were saying.

This is the excitement of growth not only in a project with a clearly articulated vision, goals, and objectives but also in an agency, where the people are clearly committed to and empowered to follow the vision of public service for which the agency is being created. Service is the ideal that guides those public servants who dare to dream of a better world, like stars guided the sailors of old.

Stage three: Growth, with the stimulation of its unknowns, its groping, its many false starts and surprising insights, eventually gives way to maturity in which the outcome seems assured. Again, I gained some sense of what this stage must be like while working on the "Blue Mountain Book."

During the first year and a half of struggling to put the ideas coherently on paper, we had many false starts. Then the pieces began coming together, slowly at first and then faster and faster. The wonderful thing was that everyone from every necessary discipline seemed to materialize "out of thin air" just when we needed them most. There was an ideal mix of people that kept the interest charged and the enthusiasm crackling.

By the end of the second year, we saw where the project was going, and we knew what the product would be—not in detail, but in general form like the broad brush strokes that define the images on a painter's canvas and so give proportion and context to the vision. At that point, the project came into its maturity, and everyone seemed to know it. From then on, it was almost impossible to keep up with the requests to put on workshops explaining the concepts embodied in the "Blue Mountain Book."

Agencies reach maturity in a similar manner. They have a function to perform and, after many false starts, begin to accomplish what they were designed to do. They mature into their stated mission, and once this has been accomplished, the turning point has been reached.

Stage four: The beginning of stage four is pivotal to the life of an agency, for here is where its direction is ultimately determined. As the "Blue Mountain Book" project reached maturity, we had to recognize and accept that it was time to begin thinking about the next project. By the time the "Blue Mountain Book" was finally published, it was like an old friend and going on to something else was extremely difficult for those of us who felt the impending death of a cherished relationship. Even my working relationships with my colleagues of five wonderful years were forever altered. It was time to move on, to risk major change, to grow.

An agency reaches the same point. Having fulfilled its original charge, it must be reenvisioned, reoriented, rechartered, and revitalized to accommodate changing times and circumstance *or* be disbanded. If left solely to its own devices, senescence creeps in. This is the point at which an agency becomes dysfunctional.

The now-declining agency tries to "hang on," to live as in the past and carry that past forward into the future, a future that is suddenly "here, now"—and often contrary to the political desires of this or that presidential administration. That is when an agency becomes a self-perpetuating machine that, having "forgotten" its original charge and having outlived its purpose, becomes dysfunctional, looking out for its own survival at any cost.

It's at this juncture that intergenerational continuity and intergenerational equity seriously break down. But simply creating another bureaucracy to replace the dysfunctional one is a nonfunctional, short-term fix, because every agency seems to follow the same pattern. Having watched this scenario play itself out again and again in public, private, and nonprofit sectors, I have come to the conclusion that some kind of politically free oversight is necessary if public forests (indeed, public lands in general) are to be cared for in perpetuity as a biological living trust. Over the years, my experience with people who dream of an ideal and dare to strive for it has led me to believe the answer to this dilemma might rest with the formation of two very special advisory councils that will—together— guide the caretaking of our public forests through the generations.

The Advisory Councils

In 1854, Abraham Lincoln said: "No man is good enough to govern another man without that other's consent." I believe that to be as true today as when Lincoln first said it. On that basis, I recommend that two advisory councils be convened. One would be a paid, full-time, nonpartisan "Forest Advisory Council" to give the professional trustees (= federal caretakers and scientists) guidance, keep them abreast of new information, and act as a reality check for their ideas. The other would be a "Children's Advisory Council" of beneficiaries from grades one through ten to inform the Forest Advisory Council about the importance of public forests to and for the generations of the future.

It is, I believe, the first obligation of professional caretakers to protect the biodiversity, genetic diversity, and functional diversity of the forests to ensure, as much as humanly possible, a biologically healthy and sustainable forest for all generations. This obligation includes protecting the ecological fertility and physical stability of the soil. The second obligation is to protect the storage and quality of the water that must ultimately be used by people in communities, towns, and cities outside of the forest boundary. And the third obligation is to maintain, insofar as biologically possible, a sustainable supply of quality wood for human use, as well as non-timber forest products.

This said, it must be recognized and accepted by members of the Forest Advisory Council that whether and how to log an area is only a choice, but a crucial one. It is a tradeoff between short-term economics and the long-term health and integrity of the forest's ecological relationships, as well as a choice of the necessities and desires of the present generations balanced against those of the future.

The Forest Advisory Council

The Forest Advisory Council (hereafter referred to as the "Forest Council") would be a national group of dedicated people willing to commit themselves to seeking ways of maintaining the greatest possible ecological health and sustainability of our public forests through time and of repairing the damage of past mismanagement. In this way, we humans can begin a new partnership with the forest, one based on the original meaning of the term "resource."

Whereas "resource" originally meant a reciprocal gift or relationship between humans and the Earth, the 1999 *Random House Webster's Unabridged Dictionary* defines "resource" as "a source of supply, support, or aid, especially one that can be readily drawn upon when needed." In this instance, "resource" is defined as a one-way transaction of use, rather than a two-way relationship, as in "re + source"—to renew in kind—in which case we use the forest and then we are truly the source of its renewal. And in keeping with the modern, extractive sense of the term, "resources" are defined as: (1) "the collective wealth of a country or its means of producing wealth" and (2) "money, or any property that can be converted into money; assets." Viewing Nature simply for its conversion potential into money is increasingly the economic theme of our time, one through which we transform spirited and lively mutual gifts into lifeless commodities, including ourselves—the "human resource."

If we, in today's world, are to effectively honor the original meaning of the term "resource," we must consciously plan to do so. The net effect of the Forest Council's purpose is to help humanity with such planning in the present for all generations as a society, drawing extensively and continuously on accruing knowledge *and* the lessons of cause and effect archived in history, while simultaneously recognizing the utility of forests to humanity and the necessity of using them.

Such an endeavor requires a meticulous practice of democracy, and democracy is "tough love." You have to want it badly because it's demanding, messy, and build around compromise while holding uncompromisingly to the ecological principles of Nature that are the linchpin to intergenerational equality when it comes to social-environmental choices and things of value from which to choose.

Another facet of democracy is the notion of human equality, in which all people are pledged to defend the rights of each person, and each person is pledged to defend the rights of all people—including those as yet unborn. In practice, however, the whole endeavors to protect the rights of the individuals, while the individuals are pledged to obey the *will* of the majority, which may or may not be wise or even just with respect to each person, in which case a person who is unjustly treated must use the democratic process to seek justice. Beyond this, democracy requires respect for and acceptance of, or at least tolerance of, others and an open exchange of ideas.

We, as people, must learn to listen to one another's ideas, not as points of unyielding positions to be defended but as different and valid experiences in a collective reality. While at times we must learn to agree to disagree, we must also learn to accept that, like blind people feeling the different parts of the proverbial elephant, each person is initially limited by his or her own perspective of what an elephant is and his or her perception within that perspective. When these things happen, we are engaged in the most fundamental aspects of democracy and come to conclusions and make decisions through participative talking, listening, understanding, compromising, agreeing, and *keeping* our agreements in an honorable way. For these reasons, I recommend creating a balanced, three-chambered system of democratic governance among the often-competing interests: environmental, social, and economic.

Once a person is appointed to a particular chamber, his or her vote on any matter for which a membership vote is required is weighted to ensure that no chamber has more influence over a decision than another. Actions in which a vote is required might include such things as how much timber is allowed to be "salvaged" following a large fire, based on ecological data; whether or not to allow the expansion of a large ski resort; or how to reconnect the vegetational mosaic of the landscape into an ecologically functional pattern over time.

The following is a *hypothetical* illustration of how the balance among chambers works in voting: Each chamber has 100 total votes it can cast on a particular issue. Now, let's suppose the environmental chamber has 50 members, the economic chamber has 100 members, and the social chamber has 10 members. Let's suppose further that 15 environmental chamber members vote in favor of a particular issue and 35 are opposed; no social chamber members vote in favor, 10 vote against; 70 economic chamber members vote in favor of the issue, 30 vote against.

Weighting of the votes:

Environmental:	100 votes/50 members = 2 votes per member
Social:	100 votes/10 members = 10 votes per member
Economic:	100 votes/100 members = 1 vote per member

The decision, based on weighted votes (must be a 2/3rds majority to pass on the first ballot), is shown in tabular form:

Chamber	In Favor	Against
Environmental	15 yeas x 2 votes = 30	35 nays x 2 votes = 70
Social	0 yeas x 10 votes = 0	10 nays x 10 votes = 100
Economic	70 yeas x 1 vote = 70	30 nays x 1 vote = 30
Totals	**100 weighted votes "yea"**	**200 weighted votes "nay"**

Matter is not passed by two-thirds majority.

This design is critical due to the necessity of ensuring, as much as possible, equal representation among the competing interests in order to obtain the best possible guidance, be kept abreast of and integrate new information, and have a multifaceted reality check on the ways and means of caretaking our public forests.

For the Forest Council to function, its members must trust the process because science, technology, and politics were designed to serve people, not the other way around, as too often happens. The Council is first and foremost an experiment in human dynamics and interrelationships at both the interpersonal level and the national level, because the people who serve on the Council are either its greatest integrative factor and thus its strength, or its greatest disruptive factor and so its weakness.

To ensure the Forest Council's greatest strength, it would be good to select for its members people like Steve Wright, the head of the Bonneville Power Administration in 2002, who returned an unexpected $7,500 bonus (more than five percent of his salary) because he "just didn't feel right" about accepting it when his agency was losing money. Wright said in a memo to his employees that he was transferring the money to the agency's budget because his workers had already suffered cuts in their budgets and he felt it was "my responsibility as the person in charge to make a personal sacrifice."[52] In addition to selecting honest, humble, and courageous people as members of the Forest Council, it is imperative that the Council's vision be clearly stated and that its overall framework be designed to encourage the integration of human ideas and talents through the transcendence of *all* interdisciplinary, political, and bureaucratic boundaries.

The integration of human ideas and talents is dependent on the maintenance of human dignity, cooperation, and coordination, as well as the free exchange of information within and among disciplines, as well as within and among cultural and political groups. The integration of ideas and talents is by no means confined to the Council, however, because the people who use the information produced by the Council would also be a constant source of its evaluation, re-evaluation, and creativity as they interpret and apply the data.

With this in mind, it is imperative that the Forest Council be soundly rooted in the latest science and technology, as well as social values, if the "weakest link" theory is to work. The weakest link theory states that no matter how strong a chain is, there always is a weakest link. So, as one fixes the weakest link in order to strengthen the whole chain, one finds that another link becomes the weakest. By always concentrating on the weakest link in the relationships among the environmental, social, and economic triad, the chain has the best chance of staying in excellent repair.

In accord with the "weakest link" idea and in deference to the tensions within the representative triad, it would be understood by the Forest Council that the requirements of Nature's inviolate ecological processes must have primacy in caretaking the public forests if humanity is to have a quality of life worth living. This means that a biologically sustainable forest is a prerequisite for a biologically sustainable yield (harvest). A biologically sustainable yield is a prerequisite for an economically sustainable forest industry. An economically sustainable industry is a prerequisite for a sustainable economy, which, finally, is a prerequisite for a society that is environmentally, socially, and economically sustainable.

Put another way, sound bio-economics (the economics of maintaining a healthy forest) must be practiced *before* sound industrio-economics (the economics of maintaining a healthy forest industry) can be practiced *before* sound socio-economics (the economics of maintaining a healthy society) can be practiced. And it all begins with a foundation mandated by Nature—a biologically sustainable forest. Only at our peril do we circumvent the ecological principles governing Nature's dynamics.

As every ecosystem functions entirely within the constraints imposed on it by Nature, it is the type, scale, and duration of the alterations to the system—the imposed constraints—that we need to be in concert with. And it is precisely because our worldview is our way of seeing how the world works—our overall perspective from which we interpret the world, our place in it, and the limits we choose to impose on Nature—that the Forest Council is composed of the three chambers. Our worldview can also be seen as a metaphysical window to the world, a window that cannot be accounted for on the basis of empirical evidence any more than it can be proved or disproved by argument of fact. "Metaphysical" simply means "beyond" (*meta*) the "physical" (*physic*), of which Albert Einstein said, "The more I study physics, the more I am drawn to metaphysics."

There are, in the most general terms, two worldviews: the sacred (largely grounded in cyclical, systems-thinking) and the commodity (largely grounded in linear-minded thinking). One need not be religious in the conventional sense to hold a sacred view of life that focuses on the intrinsic value of all things. As such, it gives birth to feelings of duty, protection, and love while emphasizing the values of joy, beauty, and caring that in turn erects *internal* constraints to potentially destructive human behavior against Nature. "Sacred" comes from the Latin *sacer*, and has the same root as *sanus*, "sane." A sacred view of life is therefore a sane view, one that corresponds to the Sanskrit *sat, cit, ananda*, or "being," "consciousness," and "bliss."

A commodity view of life is one interested in domination, control, and profit and seeks to "gain the world" by subjugating it to the human will. At the core of the commodity worldview are several economic seeds, such as self-interest, the dilemma of economy versus ecology, the growth/no growth tug-of-war, Rational Economic Man, and others.[53] With respect to this latter worldview, it is necessary to protect the health of the environment in the present for all generations through *external* constraints placed on destructive human behavior.

Eighteenth-century British philosopher and statesman Edmund Burke, considered the founding father of modern conservatism, understood well the need for external constraints placed on the destructive appetites of commodity-oriented people when he penned:

> Men [people] are qualified for civil liberty in exact proportion to their disposition to put moral chains upon their own appetites. . . . Society cannot exist unless a controlling power upon will and appetite be placed somewhere, and the less of it there is within, the more there must be without. It is ordained in the eternal constitution of things that men [people] of intemperate minds cannot be free. Their passions forge their fetters.[54]

The import of Burke's statement is linked to the sense of "our," as in collectively shared or "owned" in a community sense through generations. It also reinforces the aforementioned democratic notion of human equality, in which all people are pledged to defend the rights of each person, and each person is pledged to defend the rights of all people—including those as yet unborn.

Consequently, if we are going to stem the hemorrhage of trust in those entities that caretake public forests, such as the U.S. Forest Service and the Bureau of Land Management, and restore those forests to ecological health in any appreciable way, we must begin with attitudes, not facts. By this I mean an outer change always begins with an inner shift in attitude, a higher level of consciousness than that which caused a problem in the first place.

At this juncture, we must understand that, while the world functions perfectly, our perception of how it functions is imperfect. Although the collective level of our societal consciousness may be slowly rising in response to the transgressions of past management of our public forests, when weighed against and the potential dictates of future necessities, it is imperative that we turn to the children to add expediency in elevating the level of our adult consciousness. Having the humility and wisdom to seek the council of children is crucial because children are as yet unspoiled by the ambitions that often cause adults to suffer from pin-hole vision that blinds them to the negative consequences they bequeath the generations of the future through their myopic decisions and actions while engaged in the money chase.

The Children's Advisory Council

If you are wondering why anyone would convene a group of diverse, young children to give counsel to adults, the answer is simple. As a child, whatever opinion I dared to share with most adults, including my father, was summarily dismissed just because *I was a child* who obviously could never have a valid opinion about much of anything, which included what I wanted any part of the world to be like when I grew up. I was often dismissed in public, which made me feel inconsequential, rejected, and discounted out of hand, for no other reason than being a child. And because I sounded childish, I was often told, with some impatience, to "grow up," as though being a child was not okay. I find the situation much the same today in our harried society.

Knowing how this feels, I once asked a third-grade teacher, before I spoke to his class about forests, if he had ever asked the students what they wanted out of life or school, or, more importantly, what they wanted their forests to be like when they grew up. He said: "No." He then asked his eleven-year-old daughter if anyone had ever asked her what she wanted in life, and she also said: "No."

What has become clear to me over the years is that children in these United States *do not have* First Amendment Rights—the right of free speech, which includes the right to be heard. The "Children's Advisory Council" is meant to rectify that injustice by giving them the right of a voice through the right of honest representation in how they want *their* forests to be cared for.

With the right of free speech and the right to be heard in mind, I have in recent years worked with grade-school children to help me understand what kind of future they want us, today's adults, to leave for them as our bequest, our legacy. I have had them draw pictures and explain in writing what kind of forest they want and what they want it for. I have had them write essays on the kind of world they want when they are adults and have their first child.[55] I have had them make lists of the attributes they want in their future. And I have had discussions with them to find out how they see the world today. The following is a brief list of some of the things they want us, the trustees of their world, to leave for them. As the beneficiaries of our thoughts, decisions, and actions, the children want:

- people to be kind to one another
- people to be responsible for their own behavior and how it affects other people, animals, and the environment
- peace in the world
- people to enjoy and care for animals
- everyone to feel safe
- everyone to have enough food
- clean air and water
- a lot of trees
- forests to be healthy homes for animals
- forests to produce wood for homes
- forests in which to hike, camp, and play
- a place to really see the stars
- cars that get a lot of miles but don't pollute
- all children to have an education, including "homeless children"

Is there anything unreasonable in the above list? Is there anything impossible to achieve if we adults were to act in a way that was psychologically mature? After all, growing old is inevitable, but *growing up is optional.* Is there anything in the list that we ourselves would not like to experience in our daily lives? Is there anything that has not been sought through the ages? The German poet Johann von Goethe said it well: "All truly wise thoughts have been thought already thousands of times; but to make them truly ours, we must think them over again honestly, til they take root in our personal experience."

Because I find the children's desires reasonable, responsible, and just plain common sense, representing, as they inevitably do, the other-centered inclusivity with which we are all endowed before it is stolen from us by adult conditioning, I recommend the formation of a "Children's Advisory Council" (hereafter called the "Children's Council"). I would tell the children that the world needs their dreams, so they must hold fast to them because only in that way could the Children's Council speak on behalf of children—those present and those yet unborn—to help guide the caretaking of our public forests by advising the Forest Council on matters that concern today's children, matters that will be critical to the generations to come.

There is another equally important reason to form a Children's Council, and that is to keep adults in touch with the march of time and its changing circumstances and social values. I say this because three things happen to us as we age:

First: We, at our collective peril, forget what it is like to be a child with a child's infinite imagination of possibilities, hopes, and dreams. While the young belong in body, mind, and spirit to the present and the future, we adults too often cling to the past, the recollection of a time we laboriously drag with us into our perceived present. This "outdatedness" became apparent to me while I was still employed with the Bureau of Land Management, where I saw people in Washington, D.C., repeatedly make unwise decisions because they were based on circumstances as they remembered them from their time as newly emancipated, idealistic professionals in the field a decade or two earlier. This simply points out that we tend to become encrusted in our narrowly perceived "realities" of the present (realities that have often been formulated in an earlier time and different place), which means our sense of responsibility, as it migrates through time, is largely to protect the economic comfort of the status quo for the sake of our own generation, e.g., "non-declining, even flow" of timber to the mill—a concept that in practice would strip the forest of all its usable trees, at which time the "flow" would cease.

Second: We forget how to ask questions that are truly relevant in the present for the present *and* the future. If we are going to ask relevant *and* intelligent questions about the future of the public forests and our place in the scheme of things, we must understand and accept that most of the questions we ask deal with social values that cannot be answered through scientific investigation. Nevertheless, scientific inquiry can help elucidate the outcome of decisions based on these values and must be so employed. In addition, we would be wise to accept the gift of Zen and approach the caretaking of our forests with the beginner's mind of a child—a mind simply open to the wonders and mysteries of the Universe.

A beginner sees what the answers might be and knows not what they should be, whereas an expert "knows" what the answers should be and can no longer see what they might be. Children, as beginners, are free to explore and to discover and so hear questions in a different context than do adults who think themselves experts. Children hear questions in the context of multiple realities and infinite possibilities. When determining what question to ask, therefore, it is critical to listen to what the children say because they represent that which is to come. To them, all things are possible until adults with narrow minds, who have forgotten how to dream, put fences around their imaginations.

We adults, on the other hand, too often think we know what the answers should be and can no longer see what they might be. To us—whose imaginations were stifled by parents, schools, and the corporate-political instruction of what "reality" is—things have rigid limits of *impossibility*. We tend to become encrusted in a self-created prison of a single pet reality, with is self-limited possibilities. Because this adult reality is so limited in potential outcomes, it often turns into an obsession to be protected at any cost. We would do well, under this circumstance, to consider carefully what the children envision as possible in the future and what they want. The future, after all, is theirs.

That said, if we are going to ask truly wise questions about how best to caretake our forest, we must be open to multiple hypotheses and explanations and be willing to accept a challenge to our ideas in the spirit of learning, rather than as an invitation to combat. The greatest triumphs in science are not, after all, triumphs of facts but rather triumphs of new ways of seeing, thinking, perceiving, and asking questions. To caretake public forests as a biological living trust, we must learn to accept our ignorance and trust our intuition, while doubting our knowledge. This reversal of our adult training requires the help of children.

Third: We, as adults and parents, think we know what is good for our children, but we do so strictly from our increasingly busy adult perspective with all of its acquired baggage from growing up and the current pressures of facing an ever-more competitive world, one we tend to perpetuate. To break this cycle of relegating children to the background of life in the name of competitive necessity, we must invite them to participate in planning and evaluating the care we take of the public forests they will inherit. In this way, real continuity will be maintained as the baton of caretaking a biological living trust is passed from one generation of trustees to the next.

How might this work? I facilitated a visioning process for the community of Lakeview in south-central Oregon, and, as always, I insisted that young people were present as participants. After all the preliminary work was done, the moment of crafting the actual vision arrived. "This moment," wrote Dean Button, who used my work as a case study for his Ph.D. dissertation, "was singled out by several of the participants as one of the most memorable of the entire process." An official with the U.S. Forest Service put it this way:

They went through all kinds of gyrations at the meetings—identifying their mission and value systems and those kinds of things—and some of that worked and some of it was probably wasted energy—but the interesting thing is a lot of the stuff in here came from a couple of high school kids—I was at that meeting and what came out of those kids was really like a light went on to that committee! That's what we're really here for! [The young woman, a junior in high school, wrote the vision statement that the community unanimously adopted, whereas all the adults failed to craft an acceptable statement.] That was of value to me—I thought, man! That's great! When the local kids see what they want out of their landscape—and they live here—not the adults—it's the kids.[56]

"Months later," said Dean, this moment "was recalled with a mixture of incredulity and pride. Even some of the 'old-timers,' who expressed impatience with the emphasis paid to 'process,' had a noticeable shift in energy and enthusiasm when making the recollection."

To form a Children's Council as fairly and representatively as possible, it would have to have an equal number of boys and girls, and all ethnic backgrounds would have to be represented as equally as possible. Children selected for the Council would have to be from the inner city, as well as rural and suburban areas. They would come from families in all walks of life and all regions of our nation, including towns along lake and ocean shores, the mountains, prairies, deserts, the snow-covered reaches of Alaska and the Pacific islands of Hawaii.

To get the best possible representation of the children's collective voice, it would be necessary to visit different areas of our nation with the expressed purpose of interviewing children about what they want us, the trustees of their future, to leave for them. In addition, a new Children's Council would have to be appointed at intervals to keep their voices fresh and current; maximize their ability to be heard by adults who are often too busy and preoccupied to listen to them, no matter how sincerely, how wisely, or how loudly they speak; and to keep the vision of how we caretake our public forests alive and vital.

With interviews in hand and the first Children's Council formed, the children would meet with the Forest Council. Then, the Forest Council and the Children's Council would craft a collective vision of our public forests as a biological living trust for the present *and* the future toward which to build.

The Vision

Vision, or intelligent foresight, is an unusual capability in discernment or perception. A visionary, therefore, is one of those rare individuals for whom time is at once a link with the past and a telescope into the future. It's in this spirit that I recommend the selection of the wisest and most visionary people who can be found to serve as the Forest Council. The Forest Council's first task—with the full participation of the Children's Council—would be to derive and state a collective vision, including attendant goals, of a future toward which to build. For it is by studying short-term trends and projecting them over time as longer-term trends that it becomes possible to perceive potential, future ecological and social conditions (including economic) and our relationship with them,[57] based, of course, on the understanding that a vision is expressed in the negotiability of constraints.

Such a vision statement might be: "To treat our public forests as a perpetual, biological living trust in the present for all generations, while simultaneously recognizing their utility to humanity and the necessity of using them." A goal to augment the vision might be: "To maximize the infiltration, purification, and storage of water in forested water-catchments as much as ecologically possible for the ultimate use of the American people." And an objective to help achieve the goals and the vision might be: "To have a minimum, permanent road system, with the least possible impact on aquatic ecosystems, in all our public forests within 25 years from today."

The process of implementing a vision statement, its attendant goals, and the objectives necessary to achieve the goals, is equivalent to negotiating a series of obstacles or *constraints* levied (consciously or unconsciously) by people on themselves through the wisdom—or folly—of their vision. Some constraints, such as those imposed by Nature, are ecological and largely *nonnegotiable*, although they can be—and often have been—circumvented. Such circumvention in forestry has exacted an enormous cost in both environmental and human terms, especially when the progressively negative outcomes are passed to future generations, such as the collapse of timber-dependent communities because their forests have been severely overcut and are now either depleted or off limits to logging.

We would do well to ponder the fact that, while we can manipulate Nature, we cannot, without grave consequences, circumvent Nature. Even if we make the Earth uninhabitable for humans, Nature will create a new ecological norm. On the flip side, social, political, and economic constraints are negotiable. The feasibility of a vision for a sustainable future, in this case an ecologically healthy system of public forests, can be determined in a process that considers the constraints imposed by Nature and how they are accepted by society through consciously mediated human behavior.

The vision of some future, desired condition, by its very nature, elicits the singular social constraint that must be met if the terms of the vision are to be fulfilled, the fixed point around which everything else turns, like the hub of a wheel. A constraint in this case means being restricted to a given course of action or inaction and connotes something that limits or regulates personal behavior.

A vision does not, in and of itself, create a single constraint where none before existed. It cannot because everything in the world is already constrained and so defined by its relationship to everything else. Nothing is ever entirely free because every relationship is constantly changing, meaning *all things are variables* and *all variables are interdependent.*

What a vision does is determine the degree to which a particular socially chosen constraint is negotiable. A vision forces a blurring of interdisciplinary lines in its fulfillment because the power of the vision rests with the people who created it and those who are inspired by it, not those whose sole job it is to administer the bits and pieces of everyday life, as important as they may be.

What does "negotiable" mean here? It means to bargain for a different outcome. Consider the wettest years in the Pacific Northwest. According to historical weather records, 1996 was the wettest year on record. Other really wet years, beginning with 1896, include: 1904, 1937, 1968, 1971, 1983, and 1995.[58] Consider further that most changes in climate are determined by Nature and are *non-negotiable*. Can we negotiate with Nature to give us sunnier, drier winters without flooding when we deem the winters too dark and wet? Can we negotiate for more rain during winters we deem too dry? Well, we can try, but it will be to no avail. Nature does not negotiate, so some of the conditions Nature hands us are *nonnegotiable*. We cannot "cut a better deal," as it were, one that is more to our liking.

Our challenge is to learn what is negotiable and what is not. Beyond this, we must learn to accept with grace what is nonnegotiable, accept with humility what is negotiable, and be accountable for our personal behavior in either case. When we negotiate, we trade one set of behavioral freedoms for another in that we impose a particular constraint on ourselves as individuals through a vision in order to alleviate or free up some other potential constraint that will affect everyone in the future—the desired outcome of our vision.

Is there anything that we humans can negotiate among ourselves wherein Nature does not have the final word? The answer is, "Yes," but with the caveat— the outcome of our behavior must not breach one of Nature's ecological principles. We can, for instance, negotiate our self-created rules and regulations, our self-created economic theories and self-imposed practices. And it is exactly because our self-created, self-imposed theories, rules, regulations, and practices are negotiable that their degrees of negotiability are also open to those who would dicker for a better deal. A shared vision, on the other hand, determines, by its defined outcome, the degree of negotiability that can be afforded to any given self-created, self-imposed theory, rule, regulation, or practice in question. In other words, a vision determines the negotiability of a particular social constraint, and the constraints we deal with in everyday life are, in a human sense, social because they are behavioral. This includes such things as how one interprets the rights of private property, how one conducts oneself in church, or where one chooses to build one's house (e.g., in the floodplain of a river or on high ground). That said, all members of the Forest Council must understand and accept—or at least tolerate—the fact that the ecological principles governing Nature's dynamics are inviolate and must, of necessity, have primacy in all deliberations *if we are to pass a viable forest legacy to the next generation and beyond.*

Here it is important to understand that the more we link the present with the past, the more we will understand the present; the more we project the present into the future, the more we will understand the present. If we accept past and present as a cumulative collection of our understanding of a few finite points along an infinite continuum, the trend toward the future, then knowledge of the past tells us, albeit dimly, what the present is built on and what the future may be projected on. As a collection of finite points, any vision created in the present is necessarily based on current knowledge and memories of the past, a condition that means a vision projected into the future will contain errors and omissions and must be periodically reexamined and adjusted to correct these shortcomings. Such a perspective would help the Forest Council fulfill its charge.

The Forest Advisory Council's Charge

The Forest Council would engage in a two-fold process: (1) evaluate the condition of our public forests and assess the possibility, or even the probability, of restoring them to a healthy ecological function as an interactive system and thereby evaluate our social understanding of their resiliency in the light of drastic and often catastrophic human-caused changes and (2) construct a barometer to guide the conduct of our ecological responsibilities in the present for all generations.

To this end, the Forest Council would be designed to facilitate an evaluation of forestry problems that inherently cross all political and bureaucratic boundaries, such as the distribution of water, and to work toward their ecologically sound, socially sensitive resolution. The short-term goal of the Forest Council, with the aid of the Children's Council, is for us to view ourselves in the mirror of social-environmental consequences because it is necessary for us to evaluate and re-evaluate our choices within the context of our effects on our public forests.

In the long term, both Councils are for the children who will inherit Planet Earth, as well as whatever forest legacy we bequeath them. As such, the Forest Council's deliberations must be an unconditional gift; an honest appraisal; a bold, ongoing process of assessing and evaluating the ecological consequences of human activities in the public's forests; and it must have the explicit purpose of achieving an equitable allocation of available resources in an ecologically sustainable and socially acceptable manner between the present generation and those of the future.

Unfortunately, today's decisions are too often based on the badly out-dated premise of inexhaustible resources that are ecologically independent of one another, free for the taking, and economically cheap if exploited at someone else's environmental expense—either in space and/or time. Yet, mounting scientific data clearly show that all resources are finite, interdependent, and that many are reaching critical limits much faster than society has anticipated or those in power are generally willing to acknowledge. That notwithstanding, for each generation to receive from the preceding one the most viable, productive, sustainable, and resilient public forests that the preceding generation is capable of maintaining, what is known about forests must be synthesized into a "story line" the general public can understand.

Synthesis Teams

> *Any third-rate engineer or researcher can increase complexity; but it takes a certain flair of real insight to make things simple again.*
> — E.F. Schumacher

Programs to regulate human behavior in the name of environmental quality (e.g., the Environmental Protection Agency), human rights (e.g., laws governing civil rights) , and social equity (e.g., the United Nations) have been costing humanity several billion dollars annually, yet an integrated means to assess the effects of these regulatory programs over the long term does not exist. Although regulatory programs are based on our best understanding of the environment and our social necessities at the time of their development and implementation, it is critical that these programs be subjected to long-term evaluation in an effort to assess their effectiveness in achieving their stated vision (if they have one), goals, objectives, and to corroborate the ecological soundness and the social equity upon which they are based. This can be done only by integrating all of the ecological and social pieces that heretofore have been viewed in intellectual isolation from one another.

Further, the condition of the public forests must be evaluated through sound ecological and social assessments of the options that we are creating, maintaining, or foreclosing for the generations of the future, because the cumulative effects of our daily decisions will be inherited by all generations to come. These ecological conditions will shape the economic and social circumstances to which the generations of the future must somehow respond. Conversely, each generation's ability to respond to the multifaceted consequences of the preceding generation's decisions will depend directly on the sensitivity and the consciousness with which those decisions were made. In other words, we must understand and accept that it is neither scientific endeavors nor technological advances that will determine the options saved for future generations, but rather the level of our collective consciousness. That is to say, caretaking public forests must be done according to the immutable ecological principles understood through science and applied through human consciousness.

The availability of a given resource—especially one deemed renewable—is dependent not only on the forest's ability to renew that resource but also on our acceptance of our human responsibility to allow the forest to function in a manner consistent with its ability to renew the resource. Our whole social and economic condition—whether we understand it or not, whether we like it or not—is predicated on the interactive sum-total of individual, available resources. Acceptance of our responsibility to the forest and to the generations of the future is imperative, because all of humanity stands today at the crossroads between the end of a quality lifestyle and the beginning of an unrelenting struggle for basic survival. The scale is poised in the balance, and, as already said, it's neither scientific endeavors nor technological advances that will make the difference. Rather, it will be the level of our consciousness and our commitment to a common vision that will henceforth determine the outcome.

With this in mind, one of the first duties of the Forest Council, following the visioning process, would be to review what is known about the ecological function of forests, as well as what people from around the United States know and/ or think about their public forests. With data in hand, the council would pinpoint the ten or twelve most pressing scientific problems and the ten or twelve more pressing social problems professional caretakers will face in terms of fulfilling the agreed-to vision statement and its goals. Once these problems are identified (be they ecological, social, or economic), separate teams of nonpartisan "experts" would be assembled for the express purpose of synthesizing what is known about their respective subject and making recommendations concerning actions that need to be taken and research questions that need to be asked.

The Synthesis-Team concept can be an ongoing process for as long as humanity chooses to live in a quality environment. Without it, humanity's ability to assess, and thereby improve, its ecological and social behavior will be severely impaired at best and impossible at worst. Why? Because, in the end, we'll make our path through the 21st century by walking it—either in blind, deadly competitive grapple, like the 20th century, or cooperatively in a way that protects human dignity and enhances the quality of life.

As I envision it, each Synthesis Team (hereafter referred to as the "Team") would be paid, freed from other obligations, and given whatever equipment and support personnel it requires in order to expeditiously and effectively fulfill its mission, which mandates certain things to be done in a sequential manner. First, the interdependence of the various components of the forest and their implications for society must be acknowledged, understood, and directly accounted for. Second, particular attention must be paid to the integration of the ecological data within itself and with social and economic data. These points notwithstanding, most people neither integrate such data nor remember it. Ergo, a "story line" is necessary as a memorable integrating factor.

A simplistic example of a "story line" from the integration of data is the northern flying squirrel of the temperate coniferous forest that nests and reproduces in the tree tops and feeds primarily on belowground-fruiting fungi (truffles and false truffles, that it finds by odor. In so doing, it uses the forest from the tops of the trees down into the upper layers of the soil. As the squirrel eats and digests the fungi, the fungal spores are emitted in fecal pellets anywhere in the forest the squirrel happens to be when it defecates. As these fungal spores are washed into the soil by rain or melting snow, they inoculate the root tips of trees and act as viable extensions of a tree's root system by aiding the tree in its uptake of water, phosphorus, nitrogen, and other metabolites from the soil that the tree itself is incapable of extracting with any efficiency.

The fungi in the soil thus feed the crowns of the trees high in the air, and the trees, in turn, send sugars from photosynthesis in their crowns down to feed the fungi in the soil. The fungal fruiting bodies feed the northern flying squirrel, and the squirrel for its part "seeds" the floor of the forest with viable fungal spores that inoculate the trees' roots, whereupon the spores germinate, grow, and begin to feed the trees.[59, 60] The trees help to create the forest, which may be part of a catchment basin that also collects and stores water required for human consumption—drinking, irrigation, or electricity. Alternatively, the trees can be cut and converted into paper, lumber, or some other product for the material benefit of society. (For a more complete "story line," see the discussion of "Functional Diversity.")

Next, these data need to be interpreted, both qualitatively and quantitatively. Once interpretation is complete, reports must be prepared that clearly outline, for a general audience, the assessments of the current ecological situation based on the prevailing environmental circumstances projected into the foreseeable future. Finally, based on the above reports, the problems would be prioritized for action and prepared for resolution. This sequential structure is vital because the fundamental product derived through the Team process is information about the changing ecological conditions to be made available to scientists, decision-makers, and the general public.

Although there are many topics to choose from, the following might be among the initial reports one would ask for:

- What purposeful actions are necessary to caretake old-growth forests and roadless areas in order to protect their integrity as vanishing ecological blueprints of the forest ecosystem by emulating Nature's forest-maintaining disturbance regimes?
- Where are the outbreaks of forest-damaging insects and disease that have been caused by forestry practices, such as the suppression of fire? What is their extent? What is their prognosis? What might be done to control them without the use of chemicals?
- How can Nature's fire patterns be mapped so they can be used as caretaking templates to design landscapes that approximate the viability of the public forests prior to the suppression of fire so future generations can enjoy a biologically sustainable flow of products and amenities from vibrant and healthy public forests?
- With a fire pattern in hand as a template, what are the ecological principles necessary to design, through caretaking, a condition in which fire can be allowed to carry out its ecological role?
- What role does fire play in casehardening large woody debris, and what ecological processes does such wood fulfill in maintaining fertility of the forest soil?
- What is the minimum road system necessary to caretake a public forest, and how should it be designed and located to give all riparian habitats on all bodies of water maximum protection?
- How many people within the local area rely on forest products for their livelihood, and are the available products fairly allocated?
- How sustainable is the harvest from public forests, both in terms of the land's ecological capability and in terms of economic predictability?
- How can local communities participate in the ecological restoration of public forests?

Such information must be written so all users can understand, have confidence in, and apply the findings to improve the health of public forests as needed and to help guide the equitable allocation of available resources to society — present and future.

The Teams must begin their processes as soon as possible because, if one views the world in terms of commodities (as is the case with our Western industrialized society), it becomes apparent that time is both the commodity in the shortest supply with respect to protecting the quality of human life and the commodity that is most irreplaceable. Further, because of our reluctant, albeit growing, recognition of socially induced, environmental problems, the time to act is now.

The initial speed and scope of the Teams would be determined by their ability to obtain ecological and social data that are of good quality and integrated. To move too fast would ensure the collapse of the process for lack of clear focus and adequate integration of the apparently discrete pieces of data. In this regard, it is essential that the known supply of a given resource, the availability and renewability of that supply, and the social equity with which it is allocated be continually integrated so the effects and consequences, checks and balances, that constitute the whole can be better understood. Further, rushing the process could only generate inferior data that would stand neither the rigors of scientific and social scrutiny nor the test of time.

With the information generated through the Forest Council, the Children's Council, and the Teams, society could evaluate its ecological and social impacts on its public forests. Specifically, the information could be used as a credible basis for maximizing desirable direct and indirect human effects on the public forests, while simultaneously minimizing undesirable effects on the forest and, therefore, on the potential well-being of the next generation and beyond. Correctly perceived, the Teams and the Councils have the collective capability of uniting in a common cause that could transcend all political and bureaucratic boundaries to help protect the basic quality of human life. Further, the variable scales of circumstances addressed by the Teams necessitate that they have sufficient structure to withstand the rigorous march of time *and* sufficient flexibility to cross political and bureaucratic boundaries to accommodate the necessities of a rapidly changing society.

Whereas this book has so far been, of necessity, more conceptual than concrete, the rest is more concrete and practical than conceptual in the subjects it covers. In addition, the rest of the book deals with the ecological principles of caretaking our public forests and with monitoring the results.

Chapter 6
Caretaking the Forest

The most profound emotion we can experience is the sensation of the mystical. It is the sower of all true science.

— Albert Einstein

Although one obvious prerequisite of caretaking the public forests as a biological living trust is the honest commitment to do so (which includes allocating the necessary funds), the other, and just as important, is understanding the ecological principles that govern the forests. We humans do not and cannot "manage" forests because we are not in control of the ecological principles that govern them—and never will be. Rather, we treat them in some way, and they respond accordingly.

Under this circumstance, the professional caretakers must do their best to listen to and read what the forest(s) is telling them about its relative state of health and treat it accordingly. When, for example, I sit in the forest, I hear its heartbeat in its daily rhymes; I hear it sing through the wind and the birds, the insects and frogs; I hear its reflection in its silence; I smell its perfume in flower, soil, and mold; and I hear it breathe with bending bough, swaying grass, and dancing leaf.

These things I see, hear, smell, feel, and love in a forest are all at the behest of Nature's ecological principles, those inviolate, universal principles that govern the interrelationships of a forest—and Planet Earth, for that matter—as a living entity. In turn, the biological sustainability of a forest depends on the extent to which we humans honor those ecological principles and treat the forest in accord with them, or on the extent to which we, in our impatient arrogance, try to circumvent them. The ecological principles the Forest Council and professional caretakers must be concerned with include: (1) diversity, (2) the invisible present, (3) ecological back-ups, (4) biological capital *vs.* economic capital, (5) nature's inherent services, (6) patterns across the landscape, (7) fire in western forests, and (8) large wood in streams, rivers, estuaries, and oceans.

Diversity
Diversity is the quality of being different. It is also the variety of non-living and living things that comprise the staggering richness of the world and our human experience of it. Diversity comes in many forms, each a relationship that fits

precisely into every other relationship in the Universe and is constantly changing, constantly becoming something else.

Nature crafted the world we humans inherited through the principle of cause
and effect, which gave rise to the diversity of non-living matter. When the first
living cell came into being, diversity not only became limitless but also was responsible for the possibility of the extinction of life. This being the case, we must
understand, accept, and remember that we live in the biosphere sandwiched between the upper atmosphere (air) and the lower lithosphere (the crust of the Earth,
including the soil and water), and if we destroy any one, we will be the authors of
our own social demise, as well as the extinction of a vast diversity of living organisms.

We are already causing the exponential loss of biological, genetic, and functional diversity worldwide. Functional diversity refers to and is a measure of the
different kinds of interactions that can occur among the parts of a system. The
more kinds of interactive parts a system has, the more diverse are its possible
functions. Conversely, the fewer kinds of interactive parts a system has, the less
diverse are its possible functions.

By way of an example, let's consider a common object, a chair. A chair is a
chair because of its structure, which gives it a particular shape. A chair can be
characterized as a piece of furniture consisting of a seat, four legs, a back, and
often arms; it is an object designed to accommodate a sitting person. Because of
the seat, we can sit in a chair, and it is the act of sitting, the functional component
(a "seat") allowed by the structure, that makes a chair, a chair.

Suppose I remove the seat so that the supporting structure on which you
could once sit no longer exists. Now to sit, you must sit on the ground among the
legs of the "chair." By definition, when I removed the chair's seat, it is no longer a
chair since I have altered its structure and so altered its function. Consequently,
the structure of an object defines its function, and the function we desire from an
object or system defines its necessary structure, and both add to the ever-widen-
ing ripples of diversity, like a pebble dropped into a pool of quiet water.[61]

Diversity, in its array of interrelating scales across the time and space of a
given landscape, is little understood by the general public and so is subject to
much mistrust when agencies, such as the U.S. Forest Service, attempt to deal
with landscape-scale diversity. Some years ago, I was asked to conduct a work-
shop for the people of the Ouachita National Forest, headquartered in Hot Springs,
Arkansas. The problem was the public's concept of an acceptable scale of diversity across the landscape, a concept founded on the ignorance of scale and on
distrust, both of which were understandable.

For many years, people had watched, often with a feeling of enraged help-
lessness, as a large timber corporation clear-cut one section of forest after another, converting diverse forests into monocultural plantations of row-cropped
trees for the pulp market. Where the people had once seen an acre of forest with
a diversity of hardwood trees and shrubs, occasionally with a few conifers mixed
in, they were suddenly confronted with row after row of pines. As more and

more acres were clear-cut and converted to "fiber farms" for the pulp industry, the people developed a bias against what they perceived to be economic simplification of their beloved forest solely for some timber corporation's short-term monetary gains—that came at the expense of the aesthetic quality of the landscape-scale view, something they thought of as belonging to all of them.

Although the conversion from forest to single-species monoculture came in two scales, people of the area were consciously aware of only one, that of a diverse forest being converted into a simplified, economic fiber farm. What they did not see was the larger picture, a picture that would have been even more disturbing had they recognized it. As the timber corporation's employees clear-cut first a few acres and then another few acres, often leaving a few acres of standing forest in between because it was not corporate land, the corporation progressively created a homogeneous landscape as well as homogeneous fiber farms.

The people who used the Ouachita National Forest, on the other hand, were unknowingly advocating a hidden homogeneity by insisting on all possible diversity on all acres all of the time, theoretically eliminating any disturbance regimes that might create diversity on a larger scale. The public's insistence on small-scale diversity was based, as noted above, on both a lack of understanding of how the various scales of diversity nest one inside another and on a profound distrust of the exploitive model of "pulp forestry."

When, therefore, the folks of the Ouachita National Forest began to *restore* an indigenous, single-species pine forest and its simple ground cover of grass along one face of the Ozark Mountains, the public erupted with indignation because the people saw it simply as a maneuver to grow an even-aged monoculture of pine trees for the pulp industry. And, in their minds, they had *always known they could not trust the Forest Service!* That is when I was summoned.

My task was to help all people concerned to understand Nature's continuum of diversity: (1) how managing for small-scale diversity *on all acres all of the time* creates an unwelcome homogeneity of habitat across a landscape; (2) how different scales of diversity nest one inside another and in so doing create a collective landscape-scale habitat that is different than the individual habitats within a single scale of diversity; (3) that the pine/grass community the Forest Service was attempting to restore had indeed existed where the Forest Service was attempting to restore it, according to the journals of early settlers, as a fire-induced and fire-maintained ecosystem in times of pre-European settlement; and (4) the necessity of maintaining landscape-scale patterns of diversity if landscapes are to be adaptable to changing conditions and thereby maintain and/or produce areas suitable for human habitation over time.

It is critically important for people to both understand and accept multiple scales of diversity across an array of ecological conditions if the heterogeneity of habitats—and species—is to be maintained. They need to understand that diversity is mediated by such events as a falling leaf, a blown-over tree, a fire, a hurricane, a volcano, or an El Niño weather pattern. Each scale of disturbance alters—both destroys and creates—a habitat, or collective of habitats, by renegotiating the composition, structure, and function of plant communities that, in turn, create a time-space array of still different scales, dynamics, and dimensions of diversity that can be used by animals that alter plant communities, that become still different communities and habitats, and so on. Ergo, to caretake public forests on a biologically sustainable basis, it is necessary to have the best possible understanding of diversity itself, because biological, genetic, and functional diversity are in many ways the cumulative effect of diversity in all its various dimensions.

To understand what I mean, think of each of these kinds of diversity as an individual leg of an old-fashioned, three-legged milking stool. When so considered, it soon becomes apparent that if one leg (one kind of diversity) is lost, the stool will fall over. Fortunately, a considerable number of functional back-ups are built into an ecosystem in that more than one species (biological diversity passed forward through genetic diversity) can usually perform a similar function (functional diversity).

Biological Diversity

What seems important for sustenance is not so much biodiversity as such, but potential biodiversity, the capacity of a healthy system to respond through diversification when the need arises.

— James Lovelock, British author

Where along the continuum of naturalness a particular forest is situated depends on the ecological integrity of its biological composition, structure, and function. The problem with the myriad relationships among composition, structure, and function is that people perceive objects by means of their obvious structures or functions but seldom understand the role composition plays in either. For example, a timber company clear-cuts a forest and thereby changes the plant community's composition from predominantly trees to all herbaceous plants. This shift in the composition alters how the community functions because a dead stalk of grass is clearly not substitutable for a large dead tree when it comes to suitable habitat for a woodpecker.

Composition, if you remember from the Introduction, is the act of combining a variety of parts or elements to form a whole. In the case of a forest, composition refers to the aboveground kinds of plants that compose its basic living parts. Structure is the configuration of those parts or elements, be it simple or complex. Structure can be thought of as the organization or arrangement of the plants that form the living foundation of the forest. Function, on the other hand, is what a particular structure either can do or allows to be done within the forest. So, the composition of a forest creates its structure, and the structure determines how the forest functions. Conversely, how the forest functions dictates its necessary structure that, in turn, dictates the necessary composition.

To maintain ecological function means that the characteristics of the ecosystem must be maintained in such a way that its processes are sustainable. In other words, if you want large woodpeckers to live in your forest, you must have species of trees that not only grow large enough to accommodate them but also *are allowed* to grow large enough to accommodate them. Consequently, the characteristics of concern (as I said in the Introduction) are: (1) composition, (2) structure, (3) function, and (4) Nature's disturbance regimes that periodically alter an ecosystem's composition, structure, and function in some dramatic way.

Nature's disturbance regimes, such as fires, floods, and windstorms, tend to be environmental constraints in that they control how succession proceeds and thereby alters habitat. True, we humans can tinker with disturbance regimes, such as the suppression of fire, but in the end our tinkering catches up with us and we pay the price—such as the "fire storms" that burned millions of acres of our national forests throughout the western United States in 2002 and 2003.[62]

In addition to tinkering with a disturbance regime, we can change the trajectory of an ecosystem, such as a forest, by altering the kinds and arrangement of plants within it through "management" practices because that composition is malleable to human desire and, being malleable, is negotiable within the context of cause and effect. At this juncture, it must be understood that composition is the determiner of the structure and function in that composition is *the cause* of the structure and function, rather than its effect.

By negotiating the composition, we concurrently negotiate both the structure and function. Once the composition is ensconced, structure and function are set on a predetermined trajectory—unless, of course, the composition is drastically altered, at which time both the structure and function are altered accordingly. So, it is clearly the composition, structure, and function of a plant community that determines what kinds of animals can live there, how many, and for how long.

If we change the plant composition of a forest, we change the structure, hence the function, and that affects the animals. Under Nature's scenario, the animals are ultimately constrained by the composition because, once the composition is in place, the structure and its attendant functions operate as a unit in terms of the habitat opportunities and requirements for the various species of animals.

But then, people and Nature are continually manipulating the composition of plants, thereby changing the composition of the animals that are dependent on the structure and function of the resultant habitat. These manipulations in turn determine what uses humans can make out of the ecosystem.[63] Therefore, in order to maintain or restore the biological health of a forest so it can produce the things we valued it for in the first place, we must figure out how such a forest functions and then work backward through the required structure to the necessary composition in order to achieve that outcome. To see how this scenario might work in practice, let's consider those forest insects that are deemed "pests" because they can disrupt the economic predictability of the expected timber harvest.

Before discussing "pest" insects in the context of forests, it needs to be understood that, since Biblical times, most insects that feed in one way or another on plants we humans value for our own uses have been considered to have only negative effects on the resource, and so are thought of as "pests." On the other hand, insects are not considered pests if they feed on plants for which we find no social or economic value.

The term *pest* reflects this traditional bias and the perceived necessity of *always* having to battle them for control of the resources. Only within the past couple of decades or so has evidence emerged to show that many of the so-called "insect pests" — like all other species — enrich the world and in the process provide largely unrecognized benefits to the forest, even during apparently destructive epidemics.

Patterns of vegetation across a forested landscape, especially those created over the centuries and millennia by fire or other major disturbance regimes, influence populations of insects by: (1) establishing the spatial arrangement and degree of diversity exhibited in their sources of food, (2) controlling the quality, quantity, and distribution of the habitat for their predators, (3) influencing how insects move across a forested landscape, and (4) determining how insects alter their habitat. Not surprisingly, insects multiply and disperse much more effectively when suitable food plants are uniformly distributed; "suitable" in this sense means a given species of plant or a group of species of a certain age or size. In lush, rapidly growing, tropical rainforests, on the other hand, insect herbivory helps to create and maintain the number of available habitats.

The implications of "homogenizing" forested landscapes (simplifying a forest's composition, structure, and function) as related to insect activity are interesting and instructive. Taking a landscape of diverse, indigenous forest and homogenizing it through clear-cutting and the planting of single-species monocultural plantations has the effect of eliminating predators and physical barriers to insect dispersal, such as habitat diversity that is maintained through Nature's various interactive disturbance regimes. Loss of such habitat diversity increases both the survival of forest-damaging insects, such as bark beetles, and the likelihood of region-wide outbreaks.

To illustrate, in the Cascade Mountains of Oregon old-growth forests support far more predatory invertebrates (mainly spiders) than do "plantation forests" or even-aged stands and raises the question of how much help foresters can expect from invertebrate predators that originate in diverse, old-growth stands when it comes to controlling insects in homogeneous, even-aged stands. Although the consequences of reduced numbers and kinds of predators are impossible to predict with any certainty, it seems likely that severe epidemics of plantation-damaging insects will become increasingly frequent as diverse, indigenous forests are replaced by genetically controlled, monocultural, economically arranged "fiber farms."

How might this work? The success of plantation-damaging insects increases when the landscape is intersected with roads and managed for young, single-species monocultures of small trees. This modification of the forest makes it easier for the insects to find a suitable abundance of host trees by removing the confusion factors inherent in a diverse habitat and thereby reducing the time it takes the insects to locate food.

Simplification of the forest also reduces the diversity of habitats and the variety of species of prey necessary to maintain the populations of opportunistic predators, such as spiders and birds. These opportunistic predators are more important in preventing outbreaks of insects than are host-specific predators, such as parasitic wasps, that are dependent on finding a particular species of prey to exploit.[64]

Vegetational patterns within a forested landscape also influence the habitats available to birds and mammals that prey on defoliating insects. To clarify this point, let's examine the bird communities in coniferous forests of the Pacific Northwest because they are dominated by species that feed on insects in the foliage. Roughly 80 percent of the food consumed by northwestern birds is invertebrate prey, mostly foliage-feeding insects. It would cost about $1,800 per 1.5 square miles per year in insecticides (not counting the cost of aerial application) to kill the same number of spruce budworms (a larval moth) that are eaten by birds in the forests of north-central Washington, and this does not even count the predaceous ants that complement the birds. These insect-eating birds (such as warblers) and mammals (such as bats) depend on forests in mature to old-growth age classes for nesting and roosting. Where landscapes no longer contain the required habitat components, the numbers of these species are declining.

Hence, indigenous, mature to old-growth forests in the Pacific Northwest, with their complex array of species of both trees and predators, large size of stands, and high diversity of age classes are less vulnerable to epidemics of forest-damaging insects than are the simplified, exploitive plantations and even-aged stands. For all that, circumstances can be somewhat different in forests outside of the Pacific Northwest, such as occasional old-growth forests in the Rocky Mountains, where epicenters of insects build to epidemic proportions, yet the forest survives.

Can economic plantations, planted forests, and landscapes be designed in such a way that problems of forest-damaging insects are minimized? Yes, at least temporarily, but it will be neither simple nor easy because different insects respond differently to a given landscape pattern. A pattern that reduces problems with one insect may well create problems with another. Further, air pollution and the changing global climate will stress some forests and further stress some plantations. Such stress often translates into increased problems with forest-damaging insects and could, in time, change the plant-species composition, which presumably would alter the populations of insects and their predators.

In addition, global influences, beyond those mediated by climate, extend outward in unforeseen ways from a given forest. The most serious threat to insectivorous birds in the Pacific Northwest, for instance, may not be the loss of their summer habitat once they reach the Northwest, but rather the loss of their winter habitat in Central and South America due to logging, as well as slash and burn agriculture. Roughly one half of the species of insectivorous songbirds in the Pacific Northwest is migratory and spend the winters in tropical forests. The large-scale destruction of the tropical forests means lower numbers of insectivorous birds in the temperate forests, where they return for the summer to breed and rear their young and thereby help to control tree-damaging insects.[65]

In caretaking the public forests, it must be recognized and accepted that insects, including those causing damage to trees, are natural components of the forest even though they can reach epidemic proportions. As such, they are necessary to its long-term health as part of a complex of organisms that forms the forest's interactive "immune system." The immune system includes diseases, parasites, and a variety of predators, such as spiders, bugs, beetles, flies, wasps, ants, birds, and bats, to name a few.

If these complex, interdependent feedback loops among plants and animals are gradually simplified, so too will be the cultural aspects of humanity, such as the forest-related jobs that once depended on the feedback loops. The species that composed the feedback loops will be lost—and the feedback loops with them, both ecological and cultural, which is how the evolutionary process works.

Ecologically, it is neither good nor bad, right nor wrong for these changes to occur, although they may make the ecosystem less attractive and less usable by the humans who used to rely on it for their livelihoods and for products. If the professional caretakers, in light of the above, are to think about their explicit charge, the sustainability of our public forests over the next millennium, they have to think about the interrelationships of animals with plants, both in the context of one with the other and concurrently within the patterns we humans create across landscapes.

These same types of self-reinforcing feedback loops take place in all forests of the world, and they represent the same three basic elements: biological diversity, genetic diversity, and functional diversity.

Genetic Diversity

Genetic diversity is the way species adapt to change. The most important aspect of genetic diversity is that it can act as a buffer against the variability of environmental conditions, particularly in the intermediate and long term. Healthy environments can act as "shock absorbers" in the face of catastrophic disturbance. But what happens when trees are genetically selected by artificial means, through the intellectual isolation of linear-minded thinking, to satisfy short-term economic objectives of a supposedly independent variable within a dependent, interactive system?

Genetic manipulation of forest trees was born in the concept of short-term economic expediency. Of necessity, this process minimizes or discounts all long-term ecological ramifications to and within the forest as a whole, because tenable management practices for short-term profits must be based on predictable results. Genetic manipulation of "crop trees" is perceived by the timber industry as a partial panacea for the economic uncertainty of long-term financial expenditures in trying to control an unpredictable ecosystem through "management" (= regulation).

Nevertheless, by manipulating the genetics of the trees, the function of ecological processes in the entire "managed" forest, be it an economic plantation or otherwise, is being altered by changing how the individual trees function. In turn, all other connected biological functions are altered. And we do not even know what these functions are, let alone what difference they will make in the long-term health of the ecosystem, though we can make some educated guesses.

A central thrust of modern agricultural technology (the umbrella under which forestry has long been ideologically included) has been to: (1) isolate such individual organisms, such as a particular species of tree that possesses one or more "desirable, economic characteristics"; (2) "enhance" the desirable characteristics through artificial, selective breeding; (3) replicate the progeny on a massive scale; and (4) blanket the landscape with them in a gamble for increased short-term profits in the future. This type of plant is termed an "ideotype" (literally denoting an idea) because it is an ideal model of an industrialized, economic crop plant.

Any management, but particularly that focused on individual ideotypes, unavoidably changes other properties of the ecosystem in addition to productivity, the target of the change. Genetic diversity and its fundamental structural relationships—the architectural aspects of indigenous plant communities—are altered as well. Although numerous ecologists and occasional foresters have raised concerns about this for some time, these concerns have too often been viewed as grounded in unproven criticism that has little or nothing to do with competitive, industrial, economic "necessity."

Quite to the contrary, system-level properties are likely to play a pivotal role in the health of the ecosystem, and healthy, sustainable forests provide values that are no less real because they cannot be traded in the marketplace. Society will continue to demand wood—and water—and forests will provide it, but forests also play a central role in the dynamics of global climate, in addition to which they harbor immense, untapped biotic diversity, such as medicinal plants.

In this sense, a natural forest can be likened to a numbered Swiss bank account that has a complete denomination of its own particular currency. You do not have to see the currency in order to get the correct change. If there is a virtually unlimited amount of money in the bank in all denominations (1, 5, 10, 20, 50, and 100 in both coins and bills), then any amount chosen can be withdrawn because the collective of the denominations can be translated into the exact amount requested.

The currency in the forest (called "stored genetic variability") is unseen, as is the currency in the bank. In addition, the genetic variability in the natural forest (the possible genetic combinations) is potentially unlimited over time. This means that a forest can, within limits, adapt genetically to natural climatic changes, as well as those exacerbated by the human-caused greenhouse effect, the loss of atmospheric ozone, increasing air pollution, and so on. When genetic variability (the sum total of all denominations in the bank) is withdrawn from the forest's genetic account through such things as selective genetic manipulation for short-term economic gains, the ability of the forest to adapt to changing conditions—something it must continually do in order to survive—becomes artificially limited.

Let's take this one step further. A forest is cut down across the landscape from northern Washington to southern Oregon and from the Pacific coast to the crest of the western Cascade Mountains. "Genetically improved" Douglas fir seedlings are then planted to grow quickly. In addition, seedlings planted in northern Washington are selected to withstand cold, those in the south to withstand heat, those in the west to withstand wet weather, and those in the east to withstand dry weather. In order to artificially adapt trees to one's set of values, such as large scalr planting, certain genetic traits must be selected for. In order to do so, one must select *against* genetic flexibility, genetic plasticity—the tree's inherent ability to adapt to changing conditions.

Although Nature allows for changes in climate and equips species of trees with the genetic ability to adapt and survive, what happens when a forest is converted to a plantation selected for warm conditions (and in the process is robbed of its "undesirable" or "excess" genes) when a long-term change in the climate results in cold conditions? Likewise, what happens when a forest is converted to a plantation selected for wet conditions and a long-term change in the climate results in dry conditions, etc.?

Here looms a critical concept: the past function of an ecosystem determines its present structure, and its present structure determines its potential future function. This means that structure is defined by function, and function is determined by structure! As we humans alter the genetic composition of a species, so we alter its function in time. In this way, composition, genetic plasticity, structure, and function meld to create and maintain ecological processes both in time and across space, and it is the health of the processes that in the end creates an ecosystem as we know it, including genetic stepping-stones.

As species come and go, they enrich the world with their presence, as illustrated with insect "pests." Having considered that point, it is time to contemplate the genetics of place, that is, local populations adapted to specific habitats—genetic stepping-stones. Their importance lies in our understanding that as we fragment the landscape in which we live through such things as urban sprawl, clearcut logging, or dams in rivers we are putting our fellow planetary travelers at risk, often without even realizing it.

When we fragment landscapes, both plants and animals become vulnerable to "secret extinctions"—the loss of locally adapted populations, such as those of trees that have evolved over centuries and millennia. Such a loss can be more or less permanent and may inexorably alter the habitat because other populations of the same species might prove unable to reoccupy the habitat or might not even be able to reach it due to major changes in the environment.

As a working example, let's consider the genetics of place with sugar maples that, in New England, range from sea level up to about 2,500 feet in elevation. Depending on the elevation of their habitat, populations of sugar maples differ in a number of physiological characteristics. Those at high elevations can photosynthesize much faster than populations at middle elevations. In addition, the structure of their leaves is quite different.

Leaves of sugar maples growing at high elevations are thin. That is, for the same area of a leaf, the weight of the leaf is lowest at the high elevations and highest at the middle elevations, suggesting that sugar maples at high elevations produce cheap, throwaway leaves, much as some Mediterranean rock roses produce cheap, throwaway petals, that appear one day, stay wrinkled, and are gone before dawn the next.

Trees can produce large, thin leaves at far less cost in energy than they can thick leaves. Despite this low investment in leaf tissue, the trees' rate of photosynthesis is very high. Consequently, it stands to reason that such characteristics of low-energy, throwaway leaves coupled with high rates of photosynthesis are best adapted to the short growing seasons of high elevations. Indeed, leaf-out occurs ten days later at high elevations than at middle elevations and leaf-drop is ten days earlier—a difference of nearly three weeks in the growing season.

Because sugar maples shed their leaves each autumn and produce new ones each spring, the length of the growing season is critical to the type of leaves they produce. High-elevation sugar maples are right on the edge of conditions to which the maples are suited, and the short growing season exerts a tremendous pressure in selecting for individual trees that can produce cheap, rapidly deployed, throwaway leaves that photosynthesize quickly.

Suppose a population of sugar maples was removed from the top 350 feet of its elevational range. Could it be replaced by sugar maples from middle elevations? You could physically transplant them, but they most likely would not survive because they lack the local genetics of place. Without the local adaptations necessary to survive at the top of the species' elevational range, sugar maples simply would not be part of the plant community. If the high-elevation population becomes extinct, its former environment would be less diverse, not only by one species of tree but also by all the extant visible and invisible, aboveground and belowground, interdependent species, processes, and feedback loops.[66]

So what, you might think, if one species of tree is missing from the top of some mountain. Big deal! Well, let's consider another, more drastic example, but this time the example is hypothetical, because it hasn't happened yet—but could with increased global warming. Well, you might wonder, as a momentary aside, is global warming even real? If so, what does it mean? That depends on your point of view.

The great debate over global warming is in many ways more about personal values than about science, although it often sounds objective and scientific with numbers and equations bandied about. Researchers themselves may even believe they are engaged in a bona fide, objective, scientific debate. But are they really so engaged? Not necessarily. I say this first and foremost because no person can be truly objective, and second because even a scientist is exposed during the formative years to religious and political views of his or her parents long before being exposed to scientific thought and method.

Such deeply instilled, and perhaps unconsciously held, views have a way of slipping into whatever gaps exist in one's scientific understanding—and there are always gaps that at times are more like canyons. In addition, how one person views a given set of data and how another person views the same set of data depends on whether the person viewing the data is, by nature, a linear-minded thinker or a systems-thinker.

Then there are the inevitable holes in the data, variances in how the data are collected and interpreted, uncertainties and ignorance in the assumptions wherein small changes could result in very different projections. While both sides of the debate—warming *vs.* no warming, warming as dangerous *vs.* warming as beneficial—acknowledge these limitations, when confronted with seemingly boundless uncertainties, scientists often retreat to views that conform to their religious and political assumptions that more often than not take on a characteristic of comfort due to a measure of certainty.[67]

With this as a caveat, as more data are gathered, they are, nevertheless, pointing ever-more clearly to a globe that is warming, but unevenly. By that I mean it is evidently warming more in some places than in others. An estimated 24 cubic miles of Alaskan glaciers are melting annually, leaving yesteryear's imposing mountains of ice as today's foothills. Although Alaska still contains thirteen percent of the world's glacial ice, it is melting faster than all other ice fields combined, excluding those of Greenland and Antarctica.[68] Closer to home, Glacier National Park, in Montana, which had 150 glaciers a century ago, now has only 35, and scientists expect them to be gone within thirty years. While the melting glaciers are clearly contributing to the rising level of the world's oceans, another phenomenon must be factored into this equation, namely the expansion of the existing seawater caused by a warming climate.

Glaciers grow when more snow falls in winter than melts in summer. So, what is causing Alaska's glaciers to melt? Is it because there is less snowfall in winter and summers are warmer? Whatever the reason, something has changed in the dynamics of Alaska's glaciers, something that includes the climate.[69] In addition to the melting of the Alaskan glaciers is the waning of glaciers on Mount Kenya and Mount Kilimanjaro in Africa. In fact, the ice fields on Mount Kilimanjaro have shrunk by eighty percent in the last century and—if global warming continues—will be gone within the next fifteen to twenty years.[70]

In addition, the Arctic Ocean may well be devoid of summer ice before the end of this century. The permanent ice cap over the ocean, the ice that survives the warm summer months, is melting at about nine percent per decade. Between 1978 and 2000, an area of ice encompassing 750,000 square miles, or five times the size of England, has melted. Should the Arctic Ocean become free of summer ice, both its climate and ecology will become substantially different, perhaps without polar bears that depend on the summer ice to hunt seals, which constitute their main food.[71]

For the sake of discussion, let's assume that global warming is undisputedly real and that sugar maples range geographically from Georgia to just north of the Canadian border. If the climate were to warm up rapidly by an average of three degrees, with correspondingly higher extremes in summer temperatures occurring more frequently, the sugar maples would become stressed throughout their range. In order to survive, the species would have to compensate for the increase in temperature by migrating northward in latitude and higher in altitude.

Here is an interesting regional twist to ponder. The wispy jet contrails (trails of condensation), left high in the atmosphere, spread out into cirrus-like clouds under the right atmospheric conditions. Natural cirrus clouds, thin layers of wispy water vapor that often resemble the scale of fish, trap heat that is reflected upward from the ground and, to a lesser extent, reflect some of the Sun's rays back into space. This means contrails, formed by water vapor given off in the exhaust of jet aircraft, function like artificial cirrus clouds in regional skies crowded with air traffic, where they prevent days from getting too hot by reflecting the Sun's rays and keeping nights warmer by trapping the Earth's heat—just like the greenhouse gases that are implicated in global warming. Although the effect is negligible when averaged over the globe, which is largely free of air traffic, regionally it effectively cools the daytime temperatures and warms the nighttime temperatures by about three degrees Fahrenheit. This is an important discovery because air traffic is expected to increase at about five percent per year.[72]

What might such an increase in air traffic along the Eastern Seaboard mean to the sugar maple? What happens when the temperature increases and the maples have to migrate northward in latitude?

Given the greater connectivity of Nature's landscape since the retreat of the Wisconsin Glaciation, and given sufficient time to migrate and to adapt, such a change would have been possible, but what about now? Probably not. Why? *First*, the connectivity of Nature's landscape has been severely fragmented by our cultural tinkering; this means there are large areas through which sugar maples can no longer migrate because there is simply no suitable habitat for them. They cannot, for instance, march through cities or grow in concrete and asphalt. Nor can they grow in many other once-suitable habitats, ones that have been so drastically altered that they need time apart from human activity to once again become habitable to sugar maples.

Second, too many of the locally adapted populations within the general network of the overall sugar maple population no longer exist. They have succumbed to secret extinctions. Those genetic stepping-stones of place are no longer available as "a corridor of migration."

Forests, like people, migrate. In fact, forests tend to migrate as small, symbiotic aboveground-belowground communities of plants and animals. Even a single species of tree must have its habitat requirements met and those of its symbionts, if it is to migrate.[73] *Symbiosis* is the Greek word for living together and refers to a mutually beneficial relationship of two cohabiting organisms, each of which is a "symbiont."

Third, let's assume that the current global warming is unprecedented in speed and magnitude as a warming trend since the dawn of modern humanity, some 40,000 years ago. Even if the first two conditions still favored the migration of sugar maples, the speed of climatic change might simply be too fast for the maples to adapt. After all, the maples in Georgia would have to migrate at least to New England, and the trees in New England and southern Canada would have to occupy areas that are now boreal forest and tree line in northern Canada.

Even if we set aside this hypothetical case, we must still deal with secret extinctions. When locally adapted, interactive, aboveground-belowground communities of plants and animals (often obligatorily symbiotic, at least in the case of forests) disappear, they cannot be replaced overnight—if ever. Nor can they be replaced through the myth of "management," a concept through which we give ourselves a false sense of security and power over Nature. In any case, this scenario is further compounded by functional diversity, a concept that is almost inevitably ignored in standard, exploitive forestry practices.

Functional Diversity

All things in Nature's forest are neutral when it comes to any kind of human valuation. Nature has only intrinsic value in that each component of a forest, whether a microscopic bacterium or a towering 800-year-old tree, is allowed to develop its prescribed structure, carry out its prescribed function, and interact with other components of the forest through their prescribed, interdependent processes and feedback loops. No component is more or less valuable than another; each may differ from the other in form, but all are complementary in function (*photo 9*).

Consider that most higher plants worldwide—from the northern tundra, to the desert, to the tropical rainforest—have an obligatory symbiotic relationship with certain fungi that are central to their processes of capturing nutrients from the soil. In fact, fossil evidence of the earliest terrestrial plants, as well as current molecular studies, show that plant roots and certain fungi co-evolved as symbiotic partners to form structures known as *mycorrhizae*, a word that comes from the Greek *mykes* (a fungus) and *rhiza* (a root). Mycorrhizae are generally distributed throughout present-day plant communities.

The hyphae, or individual fungal strands, each one to two cells thick, form the main structural elements of the mycorrhizal fungi. ("Hypha," singular, comes from the Greek *hyphe*, a web.) The hyphae either penetrate the cells of a plant's roots to form an "endomycorrhiza" (*endo* = mycorrhiza *inside* the root) or ensheath the root to form an "ectomycorrhizae" (*ecto* = mycorrhiza *outside* the root).

In the nutrient-deficient conditions of most forests, at least ninety percent of a tree's "feeding" roots are colonized by ectomycorrhizal fungi and results in a layer of fungal tissue or "mantle" that forms around the tree's feeding roots, thereby creating an interface between the tree's feeding roots and the soil. From this mantle, individual hypha aggregate and organize into root-like structures called "rhizomorphs" (rhizo = root + the Greek *morphe* = shape) that grow out from the feeding roots into the soil, where they act as an extension of the tree's root system. The aggregate of hyphae are also referred to as "mycelia" (singular = mycelium), which in New Latin means "made of mushrooms." A mycelium is a given mass of hyphae that forms the non-reproductive part of a fungus.

9. Old-growth forest in western Oregon. (USDA Forest Service Photograph by James M. Trappe.)

Some mycorrhizal fungi are host-specific, meaning that a given mycorrhizal fungus colonizes a single species of plant, whereas others are generalists and colonize a number of host plants. Douglas fir or birch, for example, can be colonized by many species of mycorrhizal fungi. In turn, these fungi extend from tree to tree and so form linkages among trees. The generalized nature of host compatibility ensures that almost all trees and many other plants in an undisturbed forest ecosystem, regardless of species, are interconnected by billions of miles of hyphae, organized into mycelial systems, that stem from a diverse population of mycorrhizal fungi. If, therefore, you could pull up a whole forest and gently wash the soil from the roots of the plants, you would find a mycelial net connecting the entire forest—one of the reasons an old-growth forest is so retentive of its soil-nutrient capital.[74]

Ectomycorrhizal fungi profoundly affect the forest ecosystem. Through the obligatory, symbiotic mycorrhizal association, a plant, such as Douglas fir, provides carbohydrates to its mycorrhizal symbiont, while the fungal symbiont mediates the plant's uptake of nitrogen, phosphorus, other minerals, and water. In addition, the mycorrhizal association promotes the development of fine roots; produces antibiotics, hormones, and vitamins useful to the host plant; protects the plant's roots from pathogens and environmental extremes; moderates the effects of heavy metal toxins; and promotes and maintains soil structure and the forest food web. This mycorrhizal association is expensive, however, with an estimated fifty to seventy percent of the host plant's net annual productivity being translocated to the roots of the plant and their associated mycorrhizal fungi.[75]

When access to nutrients is increasingly restricted in a forest ecosystem that is already nutrient-impoverished, mycorrhizal fungi can influence both the interactions among plants and the species composition of the plant community itself. It appears there is an additive, beneficial effect that comes with each species of mycorrhizal fungus that colonizes a given plant, which, in turn, could mean that both biodiversity and ecosystem productivity will increase with an increasing number of fungal symbionts. This scenario seems likely because experimentation has shown that as the number of fungal symbionts is increased, so too is the collective biomass of roots and shoots, as well as the species diversity of the plant community. Conversely, as a forest is disturbed through exploitive forestry practices and the use of artificial fertilizers, the function of the mycorrhizal system can be impaired.

In addition to gleaning nutrients from the soil and translocating them into the host plant, mycorrhizal fungi, along with roots of the host plant and the free-living microbial decomposers in the soil, are significant components of the global balance of carbon. Much of the carbon balance is mediated by photosynthesis and drives the respiration or "breathing" of the soil. Photosynthesis, in turn, is the synthesis of complex organic materials (especially carbohydrates from carbon dioxide, water, and inorganic salts) by using sunlight as the source of energy, with the aid of chlorophyll and its associated pigments. The "photosynthates" (= "nutriments") produced by the process of photosynthesis (sent from the green, aboveground portion of a plant to its roots and mycorrhizal symbionts), are critical in maintaining the soil's respiration, through which carbon is extracted from the soil. Although the carbon made available to the soil through photosynthesis helps to balance the loss of carbon from the soil through respiration, the production of photosynthates is mediated more by annual seasonality than by the temperature of the soil.[76]

Keeping the above in mind, let's consider the coniferous forests of the Pacific Northwest in which Douglas fir and western hemlock predominate in the old-growth canopy. Herein lives the Northern Spotted Owl, which preys on the northern flying squirrel as a staple of its diet. The flying squirrel, in turn, depends on truffles, the belowground fruiting bodies of ectomycorrhizal fungi (a phenomenon briefly mentioned earlier).[77]

Flying squirrels dine heavily on truffles. The fate of their fecal pellets varies, however, depending on where they fall. In the forest canopy, the pellets might remain and disintegrate in the tops of the trees. Or a pellet could drop to a fallen, rotting tree and inoculate the wood. On the ground, a squirrel might defecate on a disturbed area of the forest floor where a pellet could land near a feeder rootlet of a Douglas fir that may become inoculated with the mycorrhizal fungus when the spores, having been washed into the soil by rain, germinate. If environmental conditions are suitable and root tips are available for colonization, a new fungal colony may be established. Otherwise, hyphae of germinated spores may fuse with an existing fungal thallus (the non-reproductive part of the fungus) and thereby contribute and share new genetic material.[78]

As I said before, these fungi, of which the truffles are the reproductive part, depend for survival on the live forest trees to feed them sugars produced in the trees' green crowns. In turn, the fungi form extensions of a tree's root system by collecting minerals, other nutrients, and water that are vital to the tree's survival. The fungi also depend on large, rotting trees lying on and buried in the forest floor for water and the formation of humus in the soil. *Humus*, which lends soil its dark color, is the Latin word for "the ground, soil" or alternatively the New Latin word *humos*, meaning "full of earth." Further, nitrogen-fixing bacteria occur on and in the ectomycorrhiza, where they convert atmospheric nitrogen into a form that is usable by both fungus and tree.[79]

Such mycorrhizal-small mammal-tree relationships have been documented throughout forests of the U.S. (including Alaska) and Canada.[80] They are also known from forests in Argentina,[81] Europe, and Australia.[82]

To add to the overall complexity of a late-successional, indigenous forest, a live old tree eventually becomes injured and/or sickened with disease and begins to die. As previously stated, how a tree dies determines how it decomposes and reinvests its biological capital (organic material and chemical elements) back into the soil and eventually into the next forest.

A tree may die standing as a snag to crumble and fall piecemeal to the forest floor over decades, or it may fall directly to the forest floor as a whole tree. How a tree dies is important to the health of the forest because its manner of death determines the structural dynamics of its body as habitat. Structural dynamics, in turn, determine the biological-chemical diversity hidden within the tree's decomposing body as ecological processes incorporate the old tree into the soil from which the next forest must grow. What goes on inside the decomposing body of a dying or dead tree is one example of the hidden biological and functional diversity that is totally ignored in a typical economic valuation of a forest. Consequently, that trees continue to become injured, diseased, and die is critical to the long-term biological health of a forest that itself is an interactive, organic whole defined not by the pieces of its body, but rather by the interdependent functional relationships of those pieces creating the whole—the intrinsic value of each piece and its complementary function.

Regardless of how it dies, the snag and fallen tree are, as previously mentioned, only altered states of the live tree, so the live large tree must exist before there can be a large snag or large fallen tree. A basic problem inherent in understanding this scenario lies in the fact that, if asked, few of the world's leading scientists, legislators, policy-makers, or business executives could name the five classical Kingdoms of Nature (bacteria, fungi, algae, plants, and animals) or how they complement one another. This dearth of ecological literacy shows a basic lack of understanding how ecosystems and their self-reinforcing, ecological, feedback loops work, as well as a basic lack of understanding the long-term consequences of short-term political and/or economic decisions.

Finally, the various functions of the individual species I have been speaking about, when melded in the collective, form a self-reinforcing feedback loop of mutually dependent interrelationships in which the spotted owl preys on the flying squirrel; the flying squirrel eats the truffle; the fungus is closely associated with large wood on and in the forest floor for water and the formation of humus (= the fungus, spotted owl, and flying squirrel are all dependent, either directly or indirectly, on the same large decomposing wood on the forest floor); the fungus feeds the tree; the tree feeds the fungus; the squirrel inoculates the tree's roots with fungal spores that help keep the tree healthy; the tree houses the flying squirrel and the spotted owl, and on and on *ad infinitum*.

Now comes an interesting belowground twist to the story of functional diversity. It is not only species of plants and animals that will become extinct with the liquidation of a late-successional forest, so too will the old "grandparent trees." As crop after crop of young trees replace liquidated old trees, the ecological functions performed by the old trees, such as creation of the "pit-and-mound" topography on the floor of the forest with its mixing of mineral soil and organic topsoil, become extinct processes.[83] Why? Because there are no more grandparent trees to blow over; yet, "windthrow," which creates "blowdown," is an extremely important event in forests, especially in such forests as those in southeast Alaska, where fire plays a minimal role in the disturbance regime, but windthrow is critical.[84, 85]

"Pit" in pit-and-mound topography refers to the hole left in the forest floor as a tree's roots are pulled from the soil, and "mound" refers to the soil-laden mass of roots, called a "rootwad," that is suddenly thrust into the air above the forest floor as the old tree topples. The young trees that replace the grandparent trees are much smaller and different in structure; so they cannot perform the same functions in the same ways. Of all the factors that affect the soil of the forest, the roughness of the surface caused by falling grandparent trees, particularly the pit-and-mound topography, is the most striking.

Uprooted trees enrich the forest's micro-topography by creating new habitats for vegetation. They also provide opportunities for new plants to become established in the bare mineral soil of the pit and the mound. In addition, fallen trees open the canopy, allowing more light to reach the floor of the forest, that, in combination with the pits and mounds, creates and maintains a richness in the species of plants forming the understory and the herbaceous ground cover. As large trees fall, they bring with them nitrogen-rich lichens from the canopy to the forest floor, where the lichens decompose and add their nitrogen to the soil layer. With time, the fallen tree itself presents habitats that can be readily colonized by tree seedlings and other plants.

Extinction of the grandparent trees changes the entire complexion of the forest through time, just as the function of a chair is changed when the seat is removed. The "roughness" of the forest floor that, over the centuries, resulted from the cumulative addition of pits, mounds, and fallen grandparent trees will become unprecedentedly "smooth"—without pits and mounds, without large fallen trees.

Water moves differently over and through the soil of a smooth forest floor, one that is devoid of large fallen trees acting as reservoirs, storing water throughout the heat of summer, and holding soil in place on steep slopes. Gone are the huge snags and fallen trees that acted as habitats for creatures wild and free. Gone are the stumps of the grandparent trees with their belowground "plumbing systems" of roots hollowed out by rot that guided rain and melting snow deep into the soil, as I discussed in the "Introduction" and will briefly reiterate here.

This plumbing system of decomposing tree stumps and their large, deep, hollow roots frequently form interconnected, surface-to-bedrock channels that rapidly drain water from heavy rains and melting snow. As roots slowly rot away, the collapse and plugging of these channels forces more water to drain through the soil matrix, reducing its cohesion and increasing the hydraulic pressure, thereby increasing the probability of mass soil movements.

These plumbing systems cannot be replaced by the young trees of modern forests because their roots are too small and shallow to form the conduits necessary to rapidly move large quantities of water deep into the soil. Local extinctions of the ecological functions performed by grandparent trees, alive and dead, seem to pass unnoticed into the shadow-realm of bygone forests because their passing takes place in the "invisible present."

The Invisible Present

The "invisible present," as Professor John Magnuson calls it, is our inability to stand at a given point in time and see the small, seemingly innocuous effects of our actions as they accumulate over weeks, months, and years. Consider that all of us can sense change—the growing light at sunrise, the gathering wind before a thunderstorm, the changing seasons. Some people, living for a long time in one place, can see longer-term events and remember more or less snow last winter compared to the snows of other winters, or that spring seemed early in coming last year, or that it is hotter than usual this summer.

In spite of such a gift, it is an unusual person who can sense, with any degree of precision, the changes that occur over the decades of his or her life. At this scale of time we tend to think of the world in some sort of "steady state," and we typically underestimate the degree to which change has occurred. We are unable to directly sense slow changes, and we are even more limited in our abilities to interpret their relationships of cause and effect. This being the case, the subtle processes that act quietly and unobtrusively over decades reside cloaked in the "invisible present."

The invisible present, writes Magnuson, is the scale of time within which our responsibilities for our planet are most evident. "Within this time scale, ecosystems change during our lifetimes and the lifetimes of our children and our grandchildren."[86] And hidden within this scale of time are the interwoven threads of cumulative effects, lag periods, thresholds, and the various degrees of irreversibility that cause the practice of exploitive forestry to reverberate throughout the forested ecosystem.

Cumulative effects, gathering themselves in secret, suddenly become visible. By then, it is too late to retract our decisions and actions even if the outcome they cause is decidedly negative with respect to our intentions. I say "in secret" because of the pervasive problem of our inability to perceive the gradual changes that take place in the "invisible present."

So it is that the cumulative effects of our activities compound unnoticed to a point that something in the environment shifts dramatically enough for us to see it. That shift is defined by a "threshold of tolerance" in the ecosystem, beyond which the system as we knew it, suddenly, visibly becomes something else, usually something that is immediately deemed socially undesirable, such as the loss of salmon spawning habitat.

Unimpeded access to spawning and rearing areas is a requirement for all salmon. Yet, thousands of miles of spawning streams are unavailable throughout the Pacific Northwest because of improperly installed culverts at numerous points along the region's seemingly inexhaustible network of roads. These culverts were designed and their installation was supervised over many years by a variety of engineers who thought about roads—not salmon. In Oregon alone, more than 4,000 faulty culverts are preventing salmon and steelhead from accessing roughly 8,000 miles of streams with good spawning habitat, a situation that diminishes the overall population.[87]

These culverts not only prevent the salmon from reaching their spawning streams but also deprive the forest and myriad animal life of the marine-derived nitrogen, phosphorous, and other nutrients that nourished the spawned-out salmon prior to their deaths. This nutrient deficiency is one indication of an ecosystem in trouble, of a negative cumulative effect.

Here you might ask how the cumulative effects work in Nature. Well, let's suppose that a timber company is building a road system over a 10-year period. Assume that the road system will cross 100 salmon-spawning streams when it is completed. Since a road is built as needed, culverts are installed one at a time, rather than in a systematic fashion. The first culvert is properly installed and spawning salmon can easily pass through it, but the second one forms an impassable barrier to the spawning salmon. The third and fourth culverts are okay, but the fifth, sixth, and seventh again form barriers to salmon passage. Each successive culvert that is improperly installed forms a barrier to the spawning salmon in that only the passage of water—not of salmon—was taken into account during the culvert's design and installation.

The cumulative effect of each improperly installed culvert progressively eliminates more and more spawning habitat, adversely affecting the salmon population's ability to spawn and maintain its biological viability—a *negative cumulative effect* that results from thinking strictly about the interaction of road construction and water. By the time the road system is finished, the number of improperly installed culverts reaches fifty, and more than half of the spawning habitat is unavailable to the salmon. Why more than half? Because an improperly installed culvert at the mouth of a three-branched tributary used as spawning habitat has a greater effect on the availability of salmon habitat than does an improperly placed culvert part-way up one of the branches of the tributary but with properly placed culverts on each of the other two branches.

Now the salmon becomes listed, and the timber company hires a fisheries biologist to help "manage" salmon habitat. The biologist surveys the culverts and identifies those that form barriers to salmon passage. The company, in its turn, agrees to correct the improperly installed culverts. Each culvert that is properly reinstalled reopens spawning habitat to the salmon. The cumulative effect of each properly reinstalled culvert is to progressively reopen more and more potential spawning habitat, positively affecting the salmon population's potential ability to spawn and maintain its biological viability—a *positive cumulative effect* that results from thinking like a salmon *and* a road engineer as one builds and maintains roads.

In this case, the negative effect the company's original placement of the culverts had on the salmon and the salmon-forest interactions may be reversible, but such is not always the case. How might this work with salmon? If, for example, salmon are kept out of a stream system long enough, they lose the genetic "memory" of the stream, which can result in no salmon entering for many years a restored stream.

Another example might be a dairyman who wants to drain a swamp because he thinks it would make good pasture for his milk cows. So he digs a series of ditches to lead the water away from the swamp, which effectively lowers the level of the water table. In so doing, he changes the swamp into a pasture that eliminates the habitat for a rare orchid, despite the fact that the pasture will remain as such only as long as the ditches are functional.

The orchid is then added to the endangered species list and the dairyman is ordered to reclaim the swamp, so he refills the ditches. Although the water table rises accordingly in the short term and the land once again becomes swampy, it will take much longer for the life of the swamp to return. Since the swamp habitat has been set back to its beginning, and the orchid did not return, the functional interactions of the swamp will never be as they would have been had the swamp not been drained.

A more dramatic example of such an ecosystem shift, one that is totally irreversible, was the eruption of Mount St. Helens in 1980, an ecosystem shift to which the immediate reaction from the farmers in northeastern Oregon and southeastern Washington was to lament the impending doom to their farms. Yet, what they failed to realize was that they had been farming on ash-based soils for years and that such eruptions were the foundation of the soil's fertility. As a result, it was not surprising that the farmers had a bumper crop the year following the eruption.

Once an ecosystem shifts, the effect of that shift, more often than not, is difficult and costly to reverse in even a slight measure. Despite a system's degree of reversibility, it can *never* go back to its original condition, pointing to the irreversibility of change as a dynamic. Consider, for example, that salmon pick up persistent industrial pollutants known as polychlorinated biphenyls (PCBs), which they assimilate while feeding in the ocean and then carry them over vast distances back to their natal spawning areas, such as streams and lakes. When the salmon die after spawning, they deliver their toxic cargo to the aquatic sediment. When great numbers of salmon return to a particular lake, they increase the PCB content more than sevenfold.[88]

Consequently, it is necessary to understand something about the relative fragility of simplified ecosystems (agricultural fields and economic plantations of "crop trees") as opposed to the robustness of complex ones (marshes, grasslands, and forests). Fragile ecosystems can go awry in more ways and can break down more suddenly and with less warning than is likely in robust ecosystems, because fragile systems have a larger number of components with narrow tolerances than do robust ones.[89] As such, the failure of any component can disrupt the system. When a pristine ecosystem is altered for human benefit, it is made more fragile, meaning it will require more planning and maintenance to approach the stability of the original system. While sustainability means maintaining the vital functions performed by the primeval system, or some facsimile thereof, it does not mean restoring or maintaining the primeval condition itself.

To the extent that we alter ecosystems, we make them dependent on our labor to function as we want them to. If we relax our attentiveness, they regain their power of self-determined functioning, but not usually in the way we want.

Let's look at a simple example: the linear-minded thinking that underpins exploitive forestry. The more a linear-minded forester succeeds in "controlling" or "regulating" a forest, the more biologically simplified it becomes as it is converted from a forest to an economic plantation of crop trees. In turn, the more biologically simplified the economic plantation becomes, the more fragile the system as a whole becomes with respect to its internal functioning. As its fragility increases, so does the time and energy the forester must commit to maintaining the processes that were disrupted by designing that which was pleasing to the forester's economic interests.

Take "weeding"—a linear-minded forester "weeds" with herbicides to control "competing vegetation" because a "weed," by definition, is a plant growing where the forester does not want it to grow, in this case in a manicured, economically designed tree plantation. Nevertheless, the weeds in a plantation are an important source of organic material created out of sunlight, carbon dioxide, chemical elements, and water. When they die, their organic material is committed to the soil as dead plants, where it becomes part of the source of energy for the organisms in the soil that are needed to maintain the soil's health. Because the "weeds" serve a vital function, which the forester eliminates through the act of chemical "weeding," the forester must commit to supplementing Nature's remaining nutrient capital with commercial fertilizer in order to maintain the forester's desired but fragile order.

So the more intensely one tries to control (= regulate) a forest plantation, the more intensely one *must* try to control the forest plantation in order to maintain that which is desired. This is the self-reinforcing feedback loop that linear-minded foresters create, which, by way of example, is similar to what happened in ancient Greece.

Greece, flourishing under wise agricultural use during the beginning of the Iron Age, had nevertheless greatly altered its landscape, in spite of its apparently sound agricultural ethic. But all the human-caused changes, including deforestation, do not appear to have caused the collapse of the agricultural system. It was sustainable in fact, and it might have continued to be so had it not been for the effect of outside influences.

While the Greeks modified their landscape, making it fragile, their agricultural system was sustainable as long as there was a full human population to tend the terraced fields. The destruction of their agricultural system was not a consequence of the system itself, but rather of Romans raiding the Greek countryside for slaves that reduced the population of workers and left the fragile landscape increasingly untended, thereby allowing the terraces to collapse and the soil to wash into the sea.

As long as the Greeks maintained adequate cover crops that functioned to hold in place the soil as the forests had once done, their agricultural system was sustainable. Unfortunately, as the activities of Roman slavers continually reduced the Greek's working population, there came a threshold beyond which this labor-intensive agriculture simply could not be maintained, and the system collapsed.[90]

Such approaching danger normally goes undetected in the invisible present until it is too late because ecosystems operate on the basis of lag periods, meaning there is a lag between the time when the cause of a fundamental change in an ecosystem is initiated and the time when the outcome is visibly apparent. This is somewhat analogous to the incubation period in the human body between the contracting of a disease and the manifestation of its symptoms.

For a linear-minded forester, the above discussion means that to be sustainable there must be recognition and acceptance of the reality that a person has a variable degree of control over what happens on the land—but not absolute control. The more absolutely a forester tries to control a "managed" forest or plantation, the more out of control of its ever-changing ecological processes the forester ultimately becomes and the more likely the forester is to destroy Nature's back-up systems in the process.

Ecological Back-up Systems

Although ecological back-up systems are part and parcel of composition, structure, and function, I have separated them for discussion because people confuse efficiency with effectiveness and so attempt to reduce and/or eliminate back-up systems as needless "redundancy" that is inefficient, unnecessary, and uneconomical. To understand what is meant by "efficiency," "effectiveness," and "back-up systems," let's consider two examples: pine trees and the water supply of my hometown.

Pine trees cast upon the winds of fortune a prodigious amount of pollen to be blown hither and yon. I say "the winds of fortune" because it takes an inordinate amount of pollen riding the vagaries of air currents to come in contact with and fertilize enough pine seeds to keep the species viable through time. Although an extremely *inefficient* mode of pollination in that many, many more grains of pollen are produced than are used to fertilize the available pine seeds, the system is highly *effective*, as evidenced by the persistence of pine trees through the ages. And if you are wondering what happens to all the "unneeded" grains of pollen, they are eaten by a variety of organisms that benefit from an extremely rich source of nutriment. As already noted, nothing in Nature is wasted. "Waste," as people think of it, is an *economic concept*—not an ecological one.

Now let's consider the water supply of my hometown. As its computer network was being constructed, the officials of my hometown, like businesses and communities everywhere, were increasingly focused on all conceivable aspects of *efficiency* in order to eliminate as much perceived "redundancy" (= back-up systems) as possible in everyday activities because they were—and still are—seen as a waste of money. Accordingly, everything in my hometown that could be computerized, was computerized, to eliminate the *unwanted "redundancy"* of manual control, everything, that is, except the water supply, which was fortuitously overlooked (or perhaps there was a systems-thinker in the proverbial woodpile). Today, should the computer program that controls the water supply suddenly fail, we would still have water because of the unwanted back-up system of manual override. The ability to manually override a computer failure (while considered *inefficient*) is *effective* in giving my hometown the resilience to overcome a potentially disastrous circumstance and remain viable, while other communities that were more "efficient" may not be so fortunate.

With respect to ecosystems, each contains built-in back-up systems, meaning they contain more than one species that can perform similar functions. Such back-ups give an ecosystem the resilience to either resist change and/or bounce back after disturbance. Back-up systems, in the biological sense, are comprised of the various functions of different species that act as an environmental insurance policy. To maintain this insurance policy, an ecosystem requires the three kinds of diversity: biological, genetic, and functional.

As you may recall, each of these kinds of diversity can be thought of as an individual leg of an old-fashioned, three-legged milking stool because it soon becomes apparent that if one leg (one kind of diversity) is severely damaged or lost, the stool will fall over. In reality, a considerable amount of functional back-up is built into an ecosystem in that more than one species (its biological diversity passed forward through genetic diversity) can usually perform a similar function (functional diversity).

This back-up results in a stabilizing effect similar to having a six-legged milking stool, but with two legs of different kinds of wood in each of three locations. So, if one leg is removed, it initially makes no difference which one because the stool will remain standing. If a second leg is removed, its location is crucial because, should it be removed from the same place as the first, the stool will fall. If a third leg is removed, the location is even more crucial, because removal of a third leg has now pushed the system to the limits of its stability, and it is courting ecological collapse in terms of the value we placed on the system in the first place. The removal of one more piece, no matter how well intentioned, will cause the system to shift dramatically, perhaps to our long-term social detriment.

When, therefore, we humans tinker willy-nilly with an ecosystem's composition and structure to suit our short-term economic desires, we risk losing species, either locally or totally, and so reduce, first the ecosystem's biodiversity, then its genetic diversity, and finally, its functional diversity in ways we might not even imagine.[91] With decreased diversity, we lose existing choices for manipulating our environment. This loss may directly affect our long-term economic viability because the lost biodiversity can so alter an ecosystem that it is rendered incapable of producing that for which we once valued it or that for which we, or the next generation, could potentially value it again sometime in the future.

The conceptual probability of a potential, future value makes the long-term ecological wholeness and biological richness of the forest a critical measure of economic health. If we want our public forests to be able to provide that which we desire over time, we must begin now to protect them as an ecological blueprint of composition, structure, and function and a refugium of biological and genetic diversity, rather than discounting Nature's wealth in terms of the future.[92] A refugium is an island of habitat in which a species can survive and from which it can disperse when the surrounding habitat becomes suitable for it to live in. In addition, the long-term ecological wholeness and biological richness of the forest presents us with a classroom wherein we can begin learning how to re-forest (not re-tree, but *re-forest*) the many devastated acres of our public lands. To do this, we must understand something about biological capital.

Biological Capital and the Health of Soil

"We belong to a mystery that will never belong to us," says poet John Daniel, "yet it is freely given to all who desire it. Though we distance ourselves and fail to see, it is granted everywhere and all the time. It does not fail us [although we may fail it]." Soil is a part of the vast mystery to which we belong. "To forget how to dig the earth and tend the soil is to forget ourselves," wrote Mahatma Gandhi.

Many cultures have emphasized the trusteeship of the soil through religion and philosophy because it is the crucible in which the abiotic and biotic components of life are joined to form the great "placenta" of the Earth. The biblical Abraham, in his covenant with God, was instructed: "Defile not therefore the land which ye shall inhabit, wherein I dwell."[93] The Chinese philosopher Confucius saw in the Earth's thin mantle of soil the sustenance of all life and the minerals treasured by human society. A century later in Greece, Aristotle thought of soil as the central mixing pot of air, fire, and water that formed all things.

In spite of the durability of such beliefs, most people cannot grasp their profundity because the ideas are intangible on the one hand and on the other because the march toward specialization increasingly isolates us, the "modern human," from Nature and our place in it. The seeming invisibility of the soil stems from the fact that it is as common as air and so, like air, is a birthright and taken for granted.

Although soil seems "invisible" to many people, it is at the same time thought to be divisible in that one can carve it into personal boundaries of outright ownership. In reality, soil is a seamless whole, unknown in its complexity to all people.

Whether most people understand it or not, soil is important for at least six reasons. *First*, soil is the repository of life and, as such, is the stage on which the human drama and its many constructs are physically supported and played out. *Second*, soil plays a central role in the decomposition of dead organic matter and in so doing renders harmless many potential pathogens, including those of humans, thereby adding to its store of potential nutrients. *Third*, soil stores elements that, in the proper proportions and availability, can act as nutrients for the plants growing in it. *Fourth*, soil shelters seeds and provides physical support for their roots as they germinate, grow, and mature into adult plants that seed and so perpetuate the cycle. *Fifth*, soils of various kinds, acting in concert, are a critical factor in regulating the major elemental cycles of the Earth—those of carbon, nitrogen, sulfur, and so on. And *sixth*, soil both purifies and stores water.[94]

Human society is inextricably tied to the soil for reasons beyond measurable riches, for the wealth of the Earth is archived in soil, a wealth that nurtures culture even as it sustains life, as illustrated by the following quotations: "The social lesson of soil waste is that no man has the right to destroy soil even if he does own it in fee simple. The soil requires a duty of man, which we have been slow to recognize" (H. A. Wallace, 1938). "In the old Roman Empire, all roads led to Rome. In agriculture [*and forestry*] all roads lead back to the soil from which farmers [and those in the forestry profession] make their livelihood" (G. Hambrige, 1938).

Whereas both of these statements are as true now as the day they were uttered, it is the doom of people that they too soon forget. Having said this, I must acknowledge that few people know what soil really is. Soil, the very foundation of life, is a long time in the making, as will be discussed in the following sections. Sacred as soil is, it nevertheless is all too quickly profaned and depleted! Today, for example, nearly twenty percent of the vegetated surface of the Earth has been degraded by human activities.[95]

Genesis of Soil

Soil is the result of two opposing geological forces: construction and erosion. The fiery volcanism that builds mountains and the erosive powers of wind, water, and ice that work to level them are some of the agents of these opposing forces.

A "volcano" is a vent or opening in the Earth's crust through which gases and melted rock, called "molten lava," are ejected. A volcano that is ejecting gases and molten lava is said to be active; it is in creation. On the other hand, a volcano that has "run its course" of activity is said to be "extinct" because the "life" has gone out of it.

A volcano is built from within by fire and is eroded from without by wind and water and ice. It defies gravity in its growing and falls to gravity in its dying. A volcano is born and dies and is reborn as something else. This means that volcanoes, some of which form mountains, are not eternal, but, as with all mountains, come and go, are born and die, in concert with all living things.

Consider that children, flowers, and grasses are living entities that, at given times, grow almost fast enough to actually see them increase in size. Trees grow more slowly. And some rock-dwelling lichens, which are a combination of a fungus and an alga, grow just a fraction of an inch in a century—so slowly that historians use their growth to date events.

And rocks also grow, but more slowly still. At a rock's pace of growth, the history of the Egyptian pyramids is a wink and that of the Rocky Mountains a yawn.

Rocks even have a cycle of birth and death and birth again. Some geologists estimate that since the Earth was born 4.5 billion years ago, its rocks have been through ten generations.

As mountains are born and die, as pieces of continents come and go like ships at dock in a harbor, the rock formed on the floor of the sea is raised to the tops of mountains only to be eroded into soil and returned to the floor of the sea, where it once again begins its journey through the Earth. In approximately 450 million years, the rock may reappear on the surface of the continent to begin a new life.

The constructional processes, such as volcanism, sedimentation, metamorphosis, and tectonics or deformation in the Earth's crust, are the physical and chemical methods whereby discrete bodies of rock are formed, assembled, and given their physical and chemical characteristics. Igneous processes, wherein rock is melted, may produce large bodies of homogeneous rock, such as a flow of basalt or an intrusion of granite, whereas tectonic processes (those causing the deformation of the Earth's crust) would produce in the same area a large mass of heterogeneous rock.

In contrast, through weathering processes (exposure to the effects of weather), landforms are shaped by wind, water, and ice as rock is broken down physically and chemically into smaller and smaller pieces that eventually become soil. Soil, an exchange membrane between the living and nonliving components of the Earth, is dynamic and ever changing.

Derived from the mechanical and chemical breakdown of rock and organic material, soil is built up by plants that live and die in it. It is also enriched by animals that feed on plants, evacuate their bodily wastes, and eventually die, decay, and return to the soil of their origin as organic material.

Soil, the properties of which vary from place to place within a landscape, is by far the most alive and biologically diverse part of a terrestrial ecosystem, including that of forests. The processes whereby soil develops are divided into two categories of weathering, physical and chemical, both of which depend on: (1) properties of the "parent" rock, such as its physical and chemical composition;

(2) patterns of regional and local climate; and (3) the kinds of plants and animals that are available and capable of becoming established in the newly forming soil.

Physical Weathering: Physical weathering refers to the mechanical fragmentation of rock through the actions of freezing and thawing, wetting and drying, heating and cooling, or transportation by wind, water, or ice. Freezing and thawing is an important means of fragmenting rock in climates that have cycles of temperatures cold enough to freeze rock and warm enough to thaw it. Yet, to be effective, there must be cracks and/or pores in the rock for water to enter. Once inside, water expands by approximately nine percent as it freezes. If water expands in a confined space, it can exert a pressure in excess of 30,000 pounds per square inch, far exceeding the strength of rock.

I have watched this phenomenon as some of the rocks that I used to make borders around the flowerbeds of my garden became saturated with rain. Once, the north wind came blasting out of the Canadian Arctic and froze the rocks. Although I could not see them expand, when they thawed, I often found them split into several pieces. Awesome indeed is the hidden power and effect of ice.

Cycles of wetting and drying can also break down rocks, particularly those with fine grains, although the exact mechanism is not understood.

Heating and cooling is an effective process for the gradual conversion of coarse-grained rock into smaller and smaller pieces on its journey to becoming soil. The various minerals that compose a rock have different capacities to absorb heat and to expand when heated. Changes in temperature result in stresses that lead to fracturing along the boundaries between the various minerals that make up the rock or even within some kinds of minerals themselves.

The results of heating and cooling can be seen as the flakes loosened from exposed boulders following a forest fire or at times from the rocks used to ring a camp fire. I vividly remember one experience many years ago when I camped along a stream with a limestone bed. I collected rocks from the stream's bottom and constructed a ring in which to build my fire.

My fire, burning cheerily in the middle of the circle of rounded rocks, heated the water in the limestone, causing it to expand as steam. Before I knew what was happening, the limestone began exploding, sending fragments whizzing past my head into the night beyond, while others bombarded the flames. Fortunately, none of the flying projectiles found me as their target as I scurried into the darkness. As with many lessons in my life, this one took but once to learn!

Finally, there is the transportation of rock. Sand blowing along a seashore or across a desert is an example of rock being transported by wind. Rocks can also be tumbled by the swift waters of streams and rivers or by the pounding surf of the ocean. And rocks can be moved by ice in glaciers. This all results in abrasion, which is yet another way rocks are reduced into progressively smaller pieces.

Having so far dealt with the physical breakdown and transportation of rock by gravity, water, and wind, it's time to consider chemical weathering, a phenomenon also driven by the environment.

Chemical Weathering: A rock's primary mineral composition reflects the temperature, pressure, and chemical makeup during its formation, often at a high temperature deep within the Earth. At the Earth's surface, where temperatures and pressures are lower, water and various organic and inorganic acids, as well as other chemical compounds, mediate a tenuous state of ever-changing balance. As rocks adjust to the environment at the Earth's surface, the primary minerals may be transformed into secondary minerals through chemical weathering.

Minerals weather at different rates, depending on their chemical composition and crystalline structure. Small pieces of rock and small grains of mineral break down more rapidly than large ones because small ones have a much greater surface area compared to their mass than do big ones. For this reason, a particular rock may be more susceptible to physical decomposition than to chemical decomposition. Nevertheless, initial weathering, aided by rock-eating fungi, must precede the formation of soil from hard rocks. Once soil is formed, however, the intensity of chemical breakdown is generally greater in the surrounding organic matter of the soil than in the rock itself.

The Addition of Organic Material to Mineral Soil: One of the first recognizable manifestations of organic material in soil is the formation of a dark layer near the soil's surface. This organic material comes from lichens and higher plants, such as grasses and herbs, that are capable of becoming established in raw, mineral soil. In fact, their presence greatly increases the rate at which soil is formed because they not only add to the organic layer but also act as catalysts for chemical reactions.

There are distinct differences in the distribution of organic material in soils. These differences depend on climate, slope, and the type of vegetation growing on the site. As might be expected under this scenario, soils of grassland and prairie contrast distinctly with those of a forest in the way they process and distribute organic material.

In the Midwestern United States, for example, oak forest and prairie coexist in distinct patches under a similar regime of climate. Both types of vegetation have a similar amount of organic material that includes live vegetation, vegetative litter on the surface of the soil, and organic material within the soil. But in the oak forest more than half the total organic material is tied up in the trees aboveground, whereas ninety percent of the organic material in the prairie is found within the soil.

As organic material is decomposed, it passes through many forms, but in the final analysis usually ends up as carbon dioxide that is released back into the atmosphere. There are, in addition to the atmospheric releases, some relatively stable carbon compounds known as "humus," which, as I said before, lend soil its dark color.

Incorporation of organic material into the surface of the soil, where the dark layer of "topsoil" is formed, is rapid when considered in the scale of geological time but exceedingly slow when considered in the scale of a human life. Be that as it may, as organic material increases in mineral soil, so generally does plant growth.

The molecules of humus provide many important functions, such as absorbing and holding water. In fact, the cycling of carbon through soil influences both the speed with which water can infiltrate it and how long the water can be stored before plants absorb it and "transpire" (a plant's version of "perspiring") it back into the atmosphere.

Molecules of humus also act as weak acids and form structure within the soil, such as the pores, that allow microbiological activity to exist. The porous nature of the soil provides a mechanism for holding in place both water and chemicals that are required for chemical interactions. In addition, soil quite literally resembles a discrete entity that lives and breathes through a complex mix of interacting organisms—from viruses and bacteria, to fungi, to earthworms and insects, to moles, gophers, and ground squirrels.

The activities of all these organisms in concert are responsible for developing the critical properties that underlie the basic fertility, health, and productivity of soil. The complex, biologically driven functions of the soil, in which soil organisms are the regulators of most processes that translate into a soil's productivity, may require decades to a few hundred years to develop. And there are no quick fixes if soil is extensively damaged—even by something as simple as compaction—during such activities as exploitive forestry or intensive agriculture.

The compaction of soil in forests has a negative array of cascading, cumulative effects that range from decreased air and water in the soil itself to dead roots, slowed growth of plants, increased mortality of micro flora and fauna, as well as trees, all of which reduces the productive capacity of the soil. Although soil appears to be a solid substance composed of inorganic and organic matter that you can hold in the palm of your hand and roll around between your fingers, it is much more than that.

Healthy soil has spaces filled with air between the particles and chunks that comprise its matrix. These pockets of air are created by all the organisms living in the soil—from microbes to larger animals, as well as the roots of plants. Most of these organisms depend on the availability of air and water moving through the soil in order to perform their vital, ecological functions that, in concert, create and maintain the soil's health and so that of the forest. In this sense, healthy forest soil acts more like a sponge than a brick because air normally constitutes half or more of its total volume.

To understand this, fill a gallon pail with intact, forest soil. If you then compress it, you will find that at least half of the volume was air. Just as we humans require air to breathe, so does every living thing in the soil. Clearly, compacting the soil, which eliminates the air and thereby increases the soil's density, is suffocating to everything that must breathe in order to live.

Compaction of soil also reduces its ability to absorb and store water, which simulates a drought for those organisms that do survive the initial compression of their habitat, particularly in fine-textured clays and silts. Over time, compacted soil is more prone to actual drought than is healthy, friable soil.[96]

"When you walk on your land," says my friend, Richard Hart, "please be mindful that the life under foot that is working for you is very precious. Treat it with respect. It is durable, but up to a point."

Biological Infrastructure of Soil

Organisms in the soil, such as bacteria, fungi, one-celled animals called protozoa, and worms, play critical roles in maintaining its health and fertility. These organisms perform various functions in the cycling of chemicals that are required as nutrients for the growth of green plants. Some of these functions are: (1) decomposing (recycling) plant material by bacteria and fungi; (2) improving the structure of the soil by such organisms as certain mycorrhizal fungi that produce a substance called "glycoprotein" (*glomalin*) and so increase soil aggregation;[97] (3) mediating the soil's pH, a determinant of what plants and animals can live where and what chemical reactions can take place where; and (4) controlling disease-causing organisms through competition for resources and space. Without the organisms to perform these functions, the plant communities we see on the surface of the Earth (including our public forests) would not exist.

As the total productivity of an ecosystem increases, the biological diversity within the soil's food web also seems to increase *and visa versa*. The greater the number of interactions among decomposers (organisms that decompose organic material), their predators, and the predators of the predators, the more nutrients that are retained in the soil. It is only through the belowground food web in the soil that plants can obtain the nutrients necessary for their growth. Without the belowground food web, the aboveground food web—including us humans—would cease to exist.

Due to the ever-changing complexities of soil, we humans would be wise to develop the humility necessary to accept that we will never fully understand soil; only then will we have the requisite patience to protect the organisms that perform the functions through which soil is kept healthy. Soil health *cannot* be maintained through applications of commercial fertilizer.

When commercial fertilizer is applied to agricultural row crops—*and industrial tree plantations*, it disrupts the biological infrastructure of the soil. Much fertilizer may be lost as it leaches downward through the soil into the groundwater, which it then contaminates, because neither the soil nor the organisms in the soil's disrupted food web can retain all of the added chemicals, such as nitrogen.[98] Contamination of the groundwater through the use of fertilizers, as well as other commercial chemical compounds, is a major problem in the United States.[99]

In some cases, adding fertilizer even acts like a biocide, killing the organisms in the soil's food web thereby further degrading the soil. It seems much wiser to work in harmony with the soil and the organisms that govern its infrastructure because they are responsible for the processes that provide nutrients to the plants.

The development of soil in a forest depends on self-reinforcing feedback loops, wherein organisms in the soil provide the nutrients for plants to grow, and plants in turn provide the carbon—the organic material—that selects for and alters the communities of soil organisms. One influences the other, and both determine the soil's development and health.

Although, the soil food web is a prime indicator of the health of any forest ecosystem, soil processes can be disrupted by such things as decreasing bacterial or fungal activity, decreasing the biomass of bacteria or fungi, altering the ratio of fungal to bacterial biomass in a way that is inappropriate to the desired system, reducing the number and diversity of protozoa, reducing the number of nematodes (roundworms as opposed to segmented worms like earthworms), and/or altering their community structure.

A model of a soil food web, composed of interactive strands, is enlightening because it shows that there are higher-level predators in the system whose function is to prevent the predators of bacteria and fungi from becoming so abundant they alter how the system functions. In turn, these higher-level predators serve as food for still higher-level predators.

In this way, mites, springtails, predatory roundworms, and small insects are eaten by organisms that spend much of their time aboveground. Similarly, predators in the third, fourth, and fifth upper strands of the food web are eaten by spiders (such as the Pacific folding-door spider, centipedes, ants (such as the carpenter ant, and beetles; that in turn are eaten by salamanders, birds, shrews, and mice; that in turn are eaten by snakes, still other birds, weasels, and foxes, to name a few.

With this view, it stands to reason that if part of the belowground biological diversity is lost, the soil as a system will function differently and may not produce a chosen crop in a way that meets our expectations (economic or otherwise) or may even produce a plant community not to our human liking, such as a hillside of shrubs instead of commercially valuable trees. Should the predators in the soil be lost, which disrupts its governance, the mineral nitrogen in the soil may be lost that in turn may cause poor growth in the plants and so the production of few seeds. Conversely, too many predators can overuse the bacteria and fungi, resulting in slower decomposition of organic material that is needed to fuel the system of nutrient uptake by the plants. A reduction or loss in any part of the food web affects at least two strands of the web at other levels.[100]

In addition, if we, as individuals and a society, poison the soil, for whatever reason, or otherwise directly damage its delicate infrastructure by condoning the pollution of our air and water, we help to destroy the stage on which life depends—including ours! Consequently, it is imperative for our survival that we reinvest biological capital in the health of the soil.

Biological Capital vs. Economic Capital

In a business sense, one makes money (economic capital) and then takes a percentage of those earnings and reinvests them, puts them back as a cost into the maintenance of buildings and equipment, in order to continue making a profit by protecting the integrity of the initial investment over time. In a business, one reinvests economic capital after the fact, after the profits have been earned, but it is different with biological capital, the capital of all renewable natural resources.

Biological capital (intrinsic value) must be reinvested *before* the fact, before the profits are earned. In a forest, one reinvests biological capital by leaving some proportion of the merchantable trees—both alive and dead—in the forest to rot and recycle themselves into the soil and thereby replenish the fabric of the living system.

Because some linear-minded economists and many linear-minded citizens (e.g., many shareholders in large timber corporations) refuse to accept intrinsic, ecological value as "real" value (i.e., biological capital), the use and management of natural resources are too often guided only by a cost-benefit analysis of their potential economic value when converted to something else, such as trees cut into boards. This means the only value these product-oriented people can see in our linear capitalistic system is short-term specialization, a view that is killing the soils of the Earth.

Even though protection of the forest soil can be justified economically, our ultimate connection with it escapes many people. One problem is that traditional linear economics deals with short-term tangible commodities (such as fast-growing trees planted to maximize the production of wood fiber) rather than long-term intangible values (such as the prosperity of future generations). Nonetheless, when we recognize that land, labor, and capital are finite and that every system has a carrying capacity that depends on natural or artificial support, the traditional linear economic system becomes more like a cyclic biological system.

In the late 18th century, Thomas Malthus, an English, systems-thinking economist, predicted that the human population would grow faster than the soil's ability to sustain it, but agronomic advances in the last century led many shortsighted "leaders" to dismiss this idea as simplistic and overly pessimistic. Today, however, Malthusian theory seems prophetic when one considers these trends:

- continual growth in the human population
- continual poisoning of air, soil, and water through pollution
- continual overgrazing by livestock that results in expanding deserts
- expanding global deforestation
- continual loss of the soil's protective cover of vegetation
- the reality of present famines

Linear-minded managers of exploitable forests still see protection of the soil as a cost with no benefit. Those who analyze the soil by means of traditional, linear economic analyses weigh the net worth of protecting the soil only in terms of the expected short-term revenues from future harvests of timber. They ignore the fact that it is the health of the soil that produces the yields, because the standard method for computing "soil expectation values" and economically optimal crop rotations commonly assumes that the soil's productivity remains constant or increases—but *never* declines, as alluded to by Nate Wilson, the previously quoted 2002 graduate of the forestry program at the University of Georgia. In reality, reducing the productivity of the soil on marginal sites can push the expected present net worth of subsequent harvests of timber below zero.

Given this shortsighted and flawed reasoning, it is not surprising that those who "manage" the land seldom see protection of the soil's productivity as cost effective. Yet if the real effects of management practices on long-term economic yields of forest products, *including water*, could be predicted, the invisible costs associated with poor care of the soil might be viewed differently. This is to say that if we honor Nature, we can find the place where culture and Nature meet and both are more nearly sustainable, but that requires a reassessment of "waste."

The economic concept of "waste" says in effect that anything not used directly by humans is wasted, but in a biological sense there is no such thing as waste in a forest or any other ecosystem—managed or unmanaged. Consider that a tree rotting in a forest, composting as it were, is a *re*investment of Nature's biological capital in the long-term maintenance of soil productivity, and hence of the forest itself. Biological capital includes such things as organic material and biological and genetic diversity. And to *re*invest means to invest again, to put back. How might this work?

An indigenous old-growth forest has three prominent characteristics: large live trees, large standing dead trees or snags, and large fallen trees. As noted earlier, the large snags and the large fallen trees, which are only altered states of the live old tree, become part of the forest floor and are eventually incorporated into the forest soil, where myriad organisms and processes make the nutrients stored in the decomposing wood available to the living trees. Further, the changing habitats of the decomposing wood encourage nitrogen fixation to take place by free-living bacteria. (Nitrogen fixation is the conversion of atmospheric nitrogen to a form usable by living organisms, such as a tree.) These processes are all part of Nature's rollover accounting system that includes such assets as large dead trees, biological diversity, genetic diversity, and functional diversity, all of which count as reinvestments of biological capital in the growing forest.

After the indigenous forest is liquidated, we humans may be deceived by the apparently successful growth of a first commercial tree plantation because it lives off the stored, available nutrients and ongoing processes embodied in the soil of the liquidated indigenous forest. Be not deceived, however; without balancing biological withdrawals, investments, and reinvestments, biological interest *and* principal are both spent and so both biological and economic productivity must eventually decline. The unhealthy, "managed," monocultural tree plantations of Europe—biological deserts compared to their original forests—bear testimony to such shortsighted, economic folly.

Converting a mature, natural forest to a continually exploitable forest through "management" or applying fertilizer to an existing monocultural plantation of trees is not a biological *re*investment or an economic *re*investment in either the forest or the soil. They are economic investments in "crop trees!" The initial outlay of economic capital required to liquidate the inherited forest, plant seedlings on bared land, and fertilize the young stand are economic investments in the intended product. But a forest does not function on economic capital. It functions on biological capital—leaves, twigs, branches, and the decomposing wood of large fallen trees, as well as biological diversity, genetic diversity, functional diversity.

The reinvestment of such biological capital is necessary to maintain the health of the soil that, in large measure, equates to the health of the forest. The health of the forest, in turn, equates to the long-term economic health of the timber industry and so human communities.

Planting and fertilizing tree seedlings is no more of a reinvestment in the soil of the forest than is the planting and fertilizing of wheat a reinvestment in the soil of a farmer's field or planting and fertilizing of lettuce in my garden. All are investments in the next crop. As such, they are investments in a potential product, *not* in the biological sustainability of the living system that produces the products.

As a society, we fail to reinvest in maintaining the health of biological processes because we focus only on the commercial product. We fail to reinvest because we insist that ecological variables, such as the biological health of the soil, are really constant values in the economic sense, values that can be discounted and so need not be considered when it comes to investing economic capital.

Nevertheless, with all our scientific knowledge and with all our technological skills, exploitive forestry still disallows reinvestment of biological capital in the soil because such reinvestment has come to be erroneously seen as economic waste—hence the practice of "salvage" logging to prevent such "waste." Exploitive forestry, rooted in agriculture, began with the idea that forests, considered only as collections of trees, were perpetual, economic producers of wood. With such thinking, it was necessary to convert a tree into some kind of potential economic commodity before it could be assigned a value. In assigning an initial economic value to timber, the health of the soil was ignored. Even today, with our vastly greater scientific knowledge, the health of the soil is both ignored and discounted because society is in too much of a hurry to work in unison with the land, so the industrial opiate is a large infusion of commercial fertilizers, the production of which requires tapping the finite supply of fossil fuels.

If those who caretake public forests are concerned with the health of the topsoil, as they must of necessity be, they need to accept the lessons of history. One of the most distinctive historical lessons might be that the birth of agriculture caused civilizations to rise, whereas abusive agricultural practices—based on flawed, linear economic thinking—destroyed the topsoil and fostered the collapse and extinction of some of those same civilizations.

History is replete with lessons pointing out that soil, once destroyed or lost, takes many human lifetimes to replace. And yet, with all the glaring lessons of history spread before us around the world, we insist on walking the historical path of abusing the soil.

One of the first steps along the path to protecting soil fertility is to ask how various forest-caretaking practices affect long-term ecosystem productivity, particularly that of the soil. Understanding the long-term effects of caretaking practices in turn requires knowing something about what keeps the ecosystem stable and productive. With such knowledge, we can turn our often "misplaced genius," as soil scientist Dave Perry rightly calls it, to the task of maintaining the resilience of the soil's fertility. Protecting the soil's fertility is buying an ecological insurance policy for future generations.

After all, soil is a bank of both chemical elements that are potential nutrients when available in the correct proportions and water that provides the matrix for the biological processes involved in the cycling of nutrients. In fact, of the sixteen chemical elements required by life, plants obtain all but two, carbon and oxygen, from the soil.

The soil stores essential nutrients in undecomposed litter and in living tissues and recycles them from one reservoir to another at rates determined by the previously discussed complex of biological processes and climatic factors. In a forest, the loss of nutrients in undisturbed sites is small, but some nutrients are lost when timber is cut. Others may be lost through techniques used to prepare the site for planting trees, reduce the hazard of fire, or control unwanted vegetation.

Consequently, the resilience of forested sites following a disturbance, such as cutting timber, is at least partly related to the ability of the soil to retain nutrients and water and to maintain its structural and functional integrity during the period in which plants are becoming reestablished. Beyond that, the health and fertility of the soil is reflected in the growth of the forest and the quality of the timber harvested.

Let's return directly to the soil for a moment. Sigmund Freud introduced the idea that our minds contain both conscious and unconscious parts. The soil can be thought of as our caretaking unconscious, not only because we take it for granted but also because that is where we hide our toxic wastes.[101] Remember the meltdown of the nuclear reactor at Chernobyl, in the former Soviet Union, early in 1986? Well, the meltdown was not potentially as dangerous as the buried nuclear dump that blew up near Chelyabinsk in the southern Ural Mountains in late 1957 or early 1958. "The land was dead—no villages, no towns, only chimneys of destroyed homes, no cultivated fields or pastures, no herds, no people—nothing. It was like the moon for many hundreds of square kilometers, useless and unproductive for a very long time, many hundreds of years."[102]

We must remember that the soil supports all plants growing in it, as well as myriad hidden processes that are necessary for its fertility and for healthy forests. The nuclear accident is only a more drastic, faster version of the global damage we are causing to the soil through our indifference and through the insidious poisons of our "management-introduced" array of biocides, artificial fertilizers, and the toxic wastes of industry and war. We can only sustain healthy forests by learning about and planning for the stands of trees in relation to their common denominators: soil; water; air; sunlight; biological, genetic, and functional diversity; and climate.

With the preceding in mind, it is wise to consider the observation of soil scientists V. G. Carter and T. Dale who point out that civilized people have despoiled their favorable environment mainly by depleting or destroying the natural resources:

- cutting down or burning most of the usable timber from the forested hillsides and valleys (as happened in Medieval Europe)
- overgrazing and denuding the grasslands that fed their livestock (as happened in Saharan Africa, personal observation)
- killing most of the wildlife and much of the fish and other water life (as happened in Saharan Africa, personal observation)
- permitting erosion to rob their farm land of its productive topsoil
- allowing eroded soil to clog the streams and fill their reservoirs, irrigation canals, and harbors with silt
- in many cases, using or wasting most of the easily mined metals or other needed minerals

As a result, a civilization declined in the middle of its despoliation, or the people moved to new land.[103]

Soil scientist W. C. Lowdermilk addressed this point in 1939 when he wrote, "if the soil is destroyed, then our liberty of choice and action is gone, condemning this and future generations to needless privations and dangers." To rectify society's careless actions, Lowdermilk composed what has been called the "Eleventh Commandment," which demands our full and unified attention and our unconditional embrace if human society is to survive in the long term:

> Thou shalt inherit the Holy Earth as a faithful steward, conserving its resources and productivity from generation to generation. Thou shalt safeguard thy fields from soil erosion, thy living waters from drying up, thy forests from desolation, and protect thy hills from overgrazing by thy herds, that thy descendants may have abundance forever. If any shall fail in this stewardship of the land thy fruitful fields shall become sterile, stony ground and wasting gullies, and thy descendants shall decrease and live in poverty or perish from off the face of the earth.[104]

As Lowdermilk alluded, soil lies between the living (plant and animal) and nonliving components of the landscape, an exchange membrane much like the placenta through which a mother nourishes her child. This idea is so important that it bears the reiteration of something I have already said, namely: The soil, derived from rock, is built up by plants that live and die in it. It's also enriched by animals that feed on plants, void their bodily wastes, and eventually die, decay, and return to the soil as organic matter. Soil supports the plants and animals that in turn create and maintain the myriad hidden processes that translate into soil productivity.

Soil is also the stage on which the entire human drama is enacted. Destroy the soil and the forest ceases to be! Destroy the forest and the soil becomes further impoverished and erodes, which degrades water quality. If we continue to destroy the stage on which we depend for life, we will play a progressively ebbing role in a terminal tragedy of human society.

Because soil is the placenta in which the forest grows and from which it draws sustenance, special care needs be taken to nurture it. To this end, every Caretaking District would have on staff one or two soil scientists, depending on the size of the District, the complexity of the soils, and the amount of work required to restore the forest to ecological "health" as best we humans understand it. In addition, each District would have the ability to solicit the services of outside expertise should the staff soil scientist(s) request it. Such additional services would be important because our public forests must be treated as a seamless whole that contains a kaleidoscope of interacting patterns across the broad landscapes of the various physiographic regions of the United States.

Everything I have so far discussed flows together and finally coalesces in the patterns we humans create across the forested landscapes of North America. In the end, it may well be these landscape-scale patterns that determine the quality of human life in the face of global warming and the resultant shifting of ecosystems, an exploding human population, the continual loss of arable lands to urbanization and desertification, and the accelerated loss of habitats for plants and animals that perform Nature's inherent services.

Nature's Inherent Services

The inherent services performed free by Nature constitute the invisible foundation that is the wealth of every human community, as well as their economic basis of support. In this sense, Nature's services are also the wealth of everyone involved in any way with the forest and its products, be they timber-based products or nontimber-oriented products. From a cab driver in New York City to the actor in Los Angeles, to the farmer in Kansas, and the logger in Oregon, we humans rely on oceans to supply fish; forests to supply water, wood, and new medicines; streams and rivers to transport water from its source to a point where we can access it; soil to grow food; grasslands for grazing livestock, and the list go on and on.

Although we base our livelihoods on the expectation that Nature will provide these services indefinitely, despite what we do to the environment, the economic system to which we are committing our unquestioning loyalty either undervalues, discounts, or ignores these services altogether. This is but saying that Nature's services, on which we rely for everything concerning the quality of our lives, are measured poorly or not at all. For example, in March of 2002, in Iran, Anoushiravan Najafi, the country's Deputy Director of the Department of Environment, urged immediate action because the illegal use of forests in the northern region had caused them to lose "the capability to carry out their elementary task of prevention of floods and earth erosion."[105]

Perhaps the most important of Nature's services is that rendered by the uncharismatic world of the soil under our feet, a world that in many ways is "as alien as a distant planet." Although the ecological processes occurring in the top few inches of the Earth's surface are the foundation of all terrestrial life, the hidden nature of this underworld upon which we all trod makes it exceedingly difficult to understand how it functions. As upright creatures of the sunlit world, we have an incredibly distorted view of soil—the nurturing, placental nexus between the nonliving-inorganic and the living-organic components of our world, the quality of which we continually take for granted.

The ecological processes are writ small in the geography of the soil. The ecological processes occurring within a few cubic inches of soil are equal in Nature's intrinsic stature and complexity to those in two to three acres of forest or coral reef. Despite the importance of soil, however, humanity degrades it:

Physical degradation occurs through the erosive agents of wind and water, both of which are due to the abusive practices in forestry and agriculture. With respect to soil erosion, there were three major epochs. The first arose in concert with the expansion of early river-basin civilizations, as measured by the Christian calendar, in the second millennium BC in such places as the Yellow River, Indus, Tigris-Euphrates, and so on. The second epoch occurred in the sixteen to nineteenth centuries, when stronger and sharper plowshares allowed the disruption of the sod in the steppes of Eurasia, prairies of North America, and pampas of South America, as well as other places. The third epoch commenced after the end of World War II in 1945, when an expanding human population infiltrated places heretofore untouched by mechanized agriculture.

Chemical degradation occurs in three ways: (1) soils are depleted of their available nutrients when poor farmers are forced to use every scrap of organic matter just to survive and thereby continually rob the soil of organic replenishment, (2) poor irrigation practices that saturate the land, thus increasing its salinity, and (3) dumping petrochemicals into the soil through which the soil's living web is largely destroyed, a condition that "addicts" the soil to petrochemicals if the desired plants are to grow.

Desertification takes place when land is abused to the point its protective cover of vegetation is so depleted that it becomes the playground of the winds. Poor farmers are often the architects of the wind's playground by passing their suffering onto the soil and, through abuse of the soil, to every generation that follows them. When I say poor farmers, I mean those who, by force of society, till marginal lands with little means and so reap meager yields, which guarantees the perpetuation of the farmers' poverty and their abuse of the soil in a continual, negative, self-reinforcing feedback loop through the generations.

Unfortunately, most people fail to understand that soil is the most ecologically diverse and critical ecosystem on Planet Earth. Myriad, parallel, synergistic processes are constantly taking place in the soils of the World, processes that sustain life in all its dynamic complexity.[106]

Just as soil is far more than simply "dirt" underfoot, so forests are far more than merely suppliers of wood fiber. They are a major source of the oxygen we breathe. In addition, young trees absorb vast amounts of carbon dioxide and the old trees store within their wooden bodies vast amounts of carbon (carbon sequestration), both of which help to stabilize the global climate. "Carbon seques-tration" implies that atmospheric carbon dioxide is transferred into long-lived "pools" or "sinks," where it is securely stored and thus not immediately avail-able to be reemitted into the atmosphere. Nevertheless, such sequestration of car-bon by forests, or terrestrial carbon sinks in general, may well be of short-term benefit because emissions from the burning of fossil fuel far out-weigh the amount of carbon that can be absorbed and stored over time by forests and other plant communities. Emissions from the burning of fossil fuel are currently being aug-mented by melting of the permafrost in northern latitudes, which are not only killing forests but also adding to atmospheric carbon dioxide.[107]

Forests are also the main source of water for most of the people of the world, and the value of the water they produce, purify, and store over time greatly ex-ceeds the value of whatever wood fiber humans may glean from them. They sup-ply habitat for insects, birds, and bats that pollinate crops and for birds and bats that eat insects considered to be harmful to people's economic interests, such as the forest trees themselves. In addition, forests provide wild animals and plants that can be sustainably harvested. Bluntly put, habitats, both forested and non-forested, are worth much more in terms of dollars when left intact to function as healthy ecosystems over time than when converted to a one-time, short-lived prod-uct, such as timber.

Consider that a worldwide network of nature reserves—terrestrial, fresh-water aquatic, and marine—would cost about $45 billion a year to maintain, but the loss of nature's goods and services would be somewhere between $4,400 bil-lion and $5,200 billion if these habitats are lost. To arrive at such figures, values were placed on the goods and services provided by Nature, just as businesses place values on the goods and services they provide consumers. In addition, the analysis was based on five real-life examples: (1) logging of the Malaysian tropi-cal forest; (2) small-scale agriculture that is chewing up the forests of Cameroon, West Africa; (3) destruction of mangrove swamps in Thailand, with the subse-quent loss of shrimp farming; (4) drainage of Canadian marshes for agriculture; and (5) demolition of coral reefs in the Philippines by fishing with dynamite. The economic returns of exploitation, such as the sale of timber or fish, was estimated for each one, as well as the jobs that particular industry would provide. These values were then compared with the value of the long-term goods and services from a relatively pristine neighboring ecosystem.

The real worth of Nature's services and sustainable goods is either under-valued or altogether neglected because they are provided to the public as a whole. As such, Nature's services and sustainable goods are part of the global commons— the birthright of each human being. Economics, on the other hand, deals with private interests and, generally speaking, only with conventional market-place stocks, shares, and services to which clear monetary values can be affixed. Re-member the discussion on over-exploitation and government subsidies? A prime example of this dynamic is sugarcane in Florida.

Sugarcane farmers in Florida are supported by government subsidies in order to make them economically "competitive" with sugarcane farmers in neighboring Cuba. As a result, growing sugarcane is generally unprofitable *and* continues to damage the unique wetlands of the Florida Everglades. In fact, environmentally flawed subsidies, such as that paid to artificially support the unprofitable growing of sugarcane in Florida, cost governments worldwide around $950 billion annually—enough to pay for an international network of nature reserves 20 times over.[108] This scenario raises a question: Is supporting the unprofitable growing of sugarcane in Florida—at the expense of both the public through tax dollars *and* the global commons, of which the Everglades is a part—either a wise or responsible choice on the part of our government?

Because of the importance of Nature's inherent services, usually thought of as ecosystem functions, it is worthwhile to examine one worldwide service in greater detail—pollination. Eighty percent of all cultivated crops (1,330 varieties, including fruits, vegetables, coffee, and tea) are pollinated by wild and semi-wild pollinators. Between 120,000 and 200,000 species of animals perform this service.

Bees are enormously valuable to the functioning of virtually all terrestrial ecosystems and such worldwide industries as agriculture. Pollination by naturalized European honeybees is sixty to 100 times more valuable economically than is the honey they produce. In fact, the value of wild blueberry bees is so great that farmers who raise blueberries refer to them as "flying $50 bills."[109]

While more than half of the honeybee colonies in the United States have been lost within the last fifty years, 25 percent have been lost within the last five years. Widespread threats to honeybees (other than viruses and mites) and other pollinators are fragmentation and outright destruction of their habitat (hollow trees for colonies in the case of "wild" honeybees), intense exposure to pesticides, a generalized loss of nectar plants to herbicides, as well as the gradual deterioration of "nectar corridors" that provide sources of food to migrating pollinators.

In Germany, for instance, the people are so efficient at weeding their gardens that the nation's free-flying population of honeybees is rapidly declining, according to Werner Muehlen of the Westphalia-Lippe Agricultural Office. Bee populations have shrunk by 23 percent across Germany over the past decade, and wild honeybees are all but extinct in Central Europe. To save the bees, says Muehlen, "gardeners and farmers should leave at least a strip of weeds and wildflowers along the perimeter of their fields and properties to give bees a fighting chance in our increasingly pruned and . . . [sterile] world."[110]

Besides a growing lack of food, one fifth of all the losses of honeybees in the United States is due to exposure to pesticides. Wild pollinators, such as flower flies and bumblebees, are even more vulnerable to pesticides than honeybees because, unlike hives of domestic honeybees that can be picked up and moved prior to the application of a chemical spray, colonies of wild pollinators cannot be purposefully relocated. Since at least eighty percent of the world's major crops are serviced by wild pollinators and only fifteen percent by domesticated honeybees, the latter cannot be expected to fill the gap by themselves if wild pollinators are lost.

Ironically, economic valuation of products, as measured by the Gross Domestic Product, only credits—never debits—and fosters many of the practices employed in modern intensive agriculture and exploitive forestry that actually curtail the productivity of crops by reducing pollination. An example is the high level of pesticide used on cotton crops to kill bees and other insects, which reduces the annual yield in the United States by an estimated twenty percent or $400 million.[111] In addition, herbicides often kill the plants that pollinators need to sustain themselves when not pollinating crops. Finally, squeezing every last penny out of a piece of ground by killing as much unwanted vegetation as possible in the practice of exploitive forestry or by plowing the edges of fields to maximize the agricultural planting area can reduce yields by disturbing and/or removing nearby nesting and rearing habitat for pollinators.

Unfortunately, too many people are fueled by their unquestioning acceptance of current economic theory—a theory that actively designs, condones, and encourages the destructive practices. Such people simply assume that the greatest value one can derive from an ecosystem, such as a forest, is that of maximizing its productive capacity for a single commodity in the present to the exclusion of all else.

Single-commodity production, however, is usually the least profitable and least sustainable way to use a forest because single-commodity production cannot compete with the enormous value of non-timber services, such as the production of oxygen, capture and storage of water, holding soils in place, and maintaining habitat for organisms that are beneficial to the economic interests of people and the quality of life they seek. These are all foregone when the drive is to maximize a chosen commodity in the name of a desired short-term monetary profit. Ironically, the undervalued and/or discounted and/or ignored non-timber products and amenities of the forest are not only more valuable than the production of wood fiber in the short term but also more sustainable in the long term and benefit a far greater number of people through time.

One study, illustrative of alternative strategies for managing the mangrove forests of Bintuni Bay, in Indonesia, found that leaving the forests intact would be more productive than cutting them. When the non-timber uses of the mangrove forests, such as fisheries, locally used products, and the control of soil erosion, were included in the calculation, the most economically profitable strategy was to retain the forests. Maintaining healthy mangrove forests yielded $4,800 per 2.5 acres annually over time, whereas cutting the forests would yield a one-time value of $3,600 per 2.5 acres. Maintaining the forests would ensure continued local uses of the area worth $10 million per year and provide seventy percent of the local income, while protecting a fisheries worth $25 million a year.[112]

We humans can no longer assume that the services Nature inherently performs are always going to be there because the consequences of our frequently unconscious actions often affect Nature negatively in unforeseen and unpredictable ways. Yet we can be sure that the loss of individual species and their habitats through the degradation and simplification of ecosystems can, and will, impair the ability of Nature to provide the services we need to survive with any semblance of human dignity and well-being. Losses are just that—irreversible and irreplaceable.

With respect to our public forests, the ecological integrity of the landscape and Nature's services are the principle of the trust that accrues in value and cannot be discounted because these lands are held in trust for *all* generations. The future, including that of our public lands, is a birthright of every American. Moreover, interest rates based on discounting that birthright are contrived by linear-minded economists and foresters. To discount individual products on public lands, such as timber, like it is done on private lands, is out of context with the long-term, cumulative effects of Nature's self-reinforcing feedback loops and thus is not in the best use of the trust's capital.

It is precisely because the inherent services provided by Nature are irreplaceable that we must erase the concept of obsolescence from our attitudes, our thinking, and our vocabulary. "Not only does this attitude [that obsolescence is acceptable, even desirable] undermine the conservation of vanishing species," writes ecologist David Ehrenfeld, "but it distorts our perception of our own place in nature." Ehrenfeld goes on to say that we deem ourselves to be exempt from having to play by Nature's ecological rules, that we are somehow above and apart from the game of life in the grand scheme of things. While we can pretend this works for a while, we have not, contends Ehrenfeld, been given either the permission or the power to remove ourselves from the parade of life.

"To call something obsolete," quips Ehrenfeld, "boasts an omniscience we do not possess, a reckless disregard for the deep currents of history and biology, and a supremely dangerous refusal to look at the lasting scars our technology is gashing across our planet and our souls."[113] Hence, to keep such things of value as Nature's inherent services, we must shift our thinking to a paradigm of sustainability, and we must calculate the full costs of what we do in the patterns we are creating across the landscape.

Patterns Across the Landscape

The spatial and temporal connectivity of landscape patterns is an important consideration, both in dealing with our forest legacy and our public lands as a legacy forest. I say this because agriculturally oriented foresters, and the timber companies they work for, justify their linear thinking with the adage: "We plant ten trees for every one we cut." They say this as though the number of trees can somehow replace the relationship of trees to one another in time and space, as they constitute a forest. But it is not the relationship of numbers of trees planted that confers stability on ecosystems, rather it's the pattern of their relationship—as illustrated in the respective syntactical arrangements of the words: "our," "forest," and "legacy."

Let's suppose that each word in the phrase "our forest legacy" represents a separate community: (1) "Our," (2) "Forest," and (3) "Legacy." Let's suppose further that the three communities are in immediate proximity and create a common landscape. Perhaps we can in small measure simulate the interdependence of those communities that create a harmonious landscape and those that do not by seeing how many syntactically meaningful combinations arise when we put each word of "our forest legacy" in each of its three possible locations within the phrase:

our forest legacy	~~forest our legacy~~	~~legacy our forest~~
~~forest our legacy~~	our **forest** legacy	our **legacy** forest
~~forest legacy our~~	our legacy **forest**	~~forest our **legacy**~~

There are four syntactically meaningful alignments (one in the left column, two in the center column, and one in the right column). These alignments mean that four communities, through mutual cooperation and coordination, have created social-environmental harmony within their landscapes: (1) our forest legacy and (2) our legacy forest. The other five, which are strictly self-centered and self-interested, are syntactically disharmonious and would have altered their landscapes in three different patterns—(1) forest our legacy, (2) forest legacy our, and (3) legacy our forest—to the point that they authored their own demise.

Note also that three of those disharmonious communities are the same (forest our legacy), whereas two (*forest legacy our* and *legacy our forest*) are singularly different. Here, it seems that three communities affected their landscapes in the same dysfunctional way and represent the social trance of currently accepted thinking, while the other two at least tried something different, even though it did not work.

Of the four alignments that exhibit syntactically meaningful arrangements and are harmonious, there is duplication that reduces the alignments to only two word arrangements: (1) our forest legacy and (2) our legacy forest. While both arrangements are harmonious in their own right, and each creates a landscape that nurtures the communities within it, they are different landscapes because they mean different things. In other words, "our forest legacy" means something different than "our legacy forest."

This exercise illustrates that meaning—stability—is conferred by relationship, not numbers, and demonstrates that *more is not necessarily better*. Each line had the same number of words in it, but their arrangement in space either gave them meaning as a whole or it did not, regardless of the fact that each word carried the same dictionary definition with it, irrespective of its position in the arrangement. Put another way, the effectiveness of relationship, not the number of pieces, is paramount in establishing the stability of meaning. And stability in ecosystems flows from the patterns of relationship that have evolved among the various species. Ergo, a culturally oriented ecosystem, such as an exploitive "fiber farm," that fails to support these co-evolved relationships has little chance of being sustainable.

Because ecological sustainability and adaptability depend on the connectivity of the landscape (a seamless forest ecosystem), the design created in caretaking our public forests must be anchored within Nature's evolved patterns and take advantage of them if there is to be any chance of creating a quality environment in the form of a biologically sustainable forest that is ecologically adaptable, pleasing to the public's aesthetic senses, and culturally viable through time.

As trustees of public forests, the caretakers *cannot move away from* the fragmentation of habitat; they *can only move toward* the connectivity of habitat. If we, as a people, are to have an adaptable landscape with a desirable forest legacy to pass forward, we must focus on two things—and give them primacy: (1) caretake the land for sustainable connectivity and biological richness within and across a seamless forest on the public lands, and (2) protect existing biological, genetic, and functional diversity—including habitats and plant-community types—to foster the long-term sustainability of the ecological wholeness and biological richness of the patterns we create across public lands.

The spatial patterns on landscapes result from complex interactions among physical, biological, and social forces that include the caretaking of forests. Most landscapes have also been influenced by the cultural patterns created by human use, such as farm fields intermixed with the patches of forest that surround a small town. The resulting landscape is an ever-changing mosaic of unaltered and manipulated patches of habitat that vary in size, shape, and arrangement. Such human influence can, in a sense, be thought of as an ecological disturbance.

A disturbance is any relatively discrete event that disrupts the structure of a population and/or community of plants and animals or disrupts the ecosystem as a whole and thereby changes the availability of resources and/or restructures the physical environment. Cycles of ecological disturbances, ranging from small grass fires to major hurricanes, can be characterized by their distribution in space and the size of the disturbance they make, as well as their frequency, duration, intensity, severity, and predictability.[114]

In the Pacific Northwest, for example, vast areas of connected, structurally diverse forest, of which the National Forest System was once constituted, have been fragmented and rendered homogeneous by clear-cutting small, square blocks of old-growth timber; by converting these blocks into even-aged stands of nursery stock; and by leaving small, uncut, square blocks between the clear-cuts. This "staggered-setting system," as it is called, requires an extensive network of roads. So, before half the land area was cut, almost every water-catchment was penetrated by logging roads. And when half the land was cut, the whole of the National Forest System became an all-of-a-piece patchwork quilt with few, if any, forested acres in any given area large enough to support those species of plants, birds, and mammals that require interior forest as their habitat (*photo 10, page 150*).[115]

Changing a formerly diverse landscape into a cookie-cutter sameness has profound implications. The spread of such ecological disturbances of Nature as fires, floods, windstorms, and outbreaks of insects, coupled with such disturbances of human society as urbanization and pollution, are important processes in shaping a landscape. The function of these processes is influenced by the diversity of the existing landscape pattern.

10. When half the land is cut using the "staggered-setting system," the whole forest becomes so fragmented that few, if any, forested acres are large enough to support those species that require interior forest as their habitat. (USDA Forest Service Photograph by Tom Spies)

Disturbances vary in character and are often controlled by physical features and vegetational patterns. The variability of each disturbance, along with the area's previous history and its particular soil, leads to the existing vegetational mosaic.

The greatest single disturbance to an ecosystem is usually human manipulation, often in the form of our continual and systematic attempts to control the size — to minimize the scale — of the various cycles of Nature's disturbance with which the ecosystem has evolved and to which it has become adapted. Among the most obvious are the suppression of fire and the control of flooding.

As we humans struggle to minimize the scale of Nature's disturbances, we alter a system's ability to resist or to cope with the multitude of invisible stresses to which the system adapts through the existence and dynamics of the very cycles of disturbance that we attempt to control. With respect to forests, a direct result of our attempt to minimize forest fires has caused today's fires to be more intense and more extensive than in the past because of the build-up of fuels since the onset of fire suppression.[116] Many forested areas are primed for catastrophic fire. Another example is the fact that plant-damaging insects (e.g., Douglas-fir bark beetle) and diseases spread most rapidly over areas of homogeneous tree plantations that have been stressed through the removal of Nature's disturbances to which the preexisting forest was adapted and which control an area's insects and diseases.

The precise mechanisms that allow ecosystems to cope with stress vary, but one mechanism is closely tied to the genetic plasticity of its species. Hence, as an ecosystem changes and is influenced by increasing magnitudes of stresses, the replacement of a stress-sensitive species with a functionally similar but more stress-resistant species maintains the ecosystem's overall productivity. Such replacements of species—back-ups—can result only from evolution within the existing pool of genetic diversity, a proposition that means Nature's back-ups must be protected and encouraged.[117, 118] Human-introduced disturbances, on the other hand, especially fragmentation of habitat, impose stresses ecosystems are ill adapted to cope with. In fact, fragmentation of habitat is the most serious threat to biological diversity.

Not surprisingly, biogeographical studies show that "connectivity" of habitats within a landscape is of prime importance to the persistence of plants and animals in viable numbers within their respective habitats—again, a matter of ecological diversity. In this sense, the landscape must be considered a mosaic of interconnected patches of habitats that, in the collective, act as corridors or routes of travel between and among specific patches of suitable habitats, ideally in a forest that is seamless in the caretaking sense.

Whether populations of plants and animals survive in a particular landscape depends on the rate of local extinctions from a patch of habitat and on the rate that an organism can move among patches of suitable habitat. Those species living in patches of habitat that are isolated from one another as a result of habitat fragmentation are less likely to persist.[119]

Modification of the connectivity among patches of habitat strongly influences the abundance of species and their patterns of movement. The size, shape, and diversity of patches also influence the patterns of species abundance, and the shape of a patch may determine the species that can use it as habitat. The interactions among the processes of a species' dispersal and the patterns of a landscape determine the temporal dynamics of its populations. Local populations of organisms that can disperse great distances may not be as strongly affected by the spatial arrangement of patches of habitat, as are more sedentary species. For example, relatively sedentary species, such as salamanders and red tree mice, can survive in a relatively small, isolated patch of quality habitat, but would disappear if the habitat was altered in such a way that they could no longer use it. Wide-ranging species, such as elk and mountain lions, on the other hand, can travel great distances from one suitable patch of habitat to another and so are much more flexible in their use of habitat. Moreover, a species' habitat for reproduction is much more restrictive than its habitat for feeding.

The way timber-harvest units are placed on the landscape in both time and space affects the overall connectivity of the landscape patterns, for better or ill, it is wise to have a landscape-scale template to work with. I say this because the transition from Nature's forests in the Pacific Northwest to "managed," economic plantations has been marked by many changes and accompanied by grave, ecological uncertainties. Fire, Nature's primary disturbance regime, was historically a common feature of the western coniferous forests and woodlands, mostly low-intensity fires with occasional large fires of high intensity in the Cascade Mountains of Oregon and Washington.

On the other hand, in the interior forests east of the Cascade Mountains into the northern Rocky Mountains, both fire and periodic outbreaks of insects appear to be part of the normal, historic, environmental pattern to which organisms that comprise the forests were adapted.[120] In fact, some species, such as *Melanophila miranda* Lec., a wood-boring beetle in the family Buprestidae, have evolved a direct relationship with fire. Female *Melanophila miranda* in the Great Basin of the American West are so attuned to laying their eggs on burned juniper trees that they are known to land on wood so hot from fire that it burns their feet off.

These fire patterns are Nature's disturbances, and inasmuch as they represent the natural conditions that created the forests, they are a healthy part of landscape-scale diversity when viewed over time. Unlike forest fires of a century ago, however, those of today are increasingly frequent and destructive—both to forests and private property.[121] Such fires are promoted by: (1) the long history of fire suppression, (2) the build-up of dead wood that accompanied fire suppression, (3) the unabated growth of shade-tolerant understory trees that accompanied fire suppression, and (4) the continuing trend toward homogeneous monocultures of young trees.

The probability of a lightning strike in the coniferous forests of the western United States turning into a hard-to-control fire depends on the amount, flammability, and distribution of fuels—both vertically and horizontally. Shade-tolerant trees in the understory (termed a "fire ladder" or a "fuel ladder") provide a vertical avenue for the spread of a fire from the ground into the crowns of the overstory trees.[122] Once in the crowns, the degree to which the crowns are "packed" (touching one another horizontally) influences the spread of a fire across the landscape. Like insect "pests," fire spreads most rapidly where it has an abundant, contiguous, and homogeneous source of "food."

Old-growth, coniferous forests (200 years +) are frequently considered to be vulnerable to catastrophic fire because of the large amounts of fuel on the ground (particularly since the days of fire suppression) and the sometimes-abundant ladder fuels. Although older forests in the northern Rocky Mountains, particularly those in the transition zone from lodgepole pine to spruce and fir, are indeed the most susceptible to fire, the situation appears to be different in forests of southwestern British Columbia, the western Cascade Mountains of Washington and Oregon, and the western Sierra Nevada of California. These forests appear to be most susceptible to fire during their first 75 to 100 years.

Here, old-growth forests are less vulnerable to forest-replacing fire, because they frequently have a greater patchiness in the crown layer that hampers the spread of flames. This is true except during "fire years" or "super fire years" when everything is set to burn due to hot, dry, easterly winds and/or conditions of severe drought. Even though older stands have tended to develop fuel ladders in the form of understory trees since the onset of fire suppression, such an understory, depending on its flammability, may or may not act as a fuel ladder.

Fires in southwestern Oregon over the last couple of decades provide compelling evidence that stands with a component of hardwoods are less severely burned than those without. The reason is straightforward: the foliage of most hardwood shrubs and trees is not very flammable and, consequently, burns poorly, except during regional "fire years" or "super fire years."

As long as dry summers prevail in the region, fires will periodically occur, and those in culturally designed landscapes can be very destructive and exceedingly difficult to control. Furthermore, exploitive forest-management practices are continually creating fire-prone landscapes in Oregon, and throughout the Western United States, for these reasons: (1) the structure and wide distribution of young, even-aged plantations west of the Cascade Mountains has greatly increased their vulnerability to catastrophic fires because of their tightly packed crowns and their uniform ages of less than 100 years and (2) east of the Cascade Mountains, the suppression of fire for so many decades has created fuel ladders in many forested areas that now endanger the crowns of the trees.

Both of these scenarios were in play on the 21st of July 2002, when 25 major forest fires were burning in Oregon alone. They had charred 212,000 acres, and some were still out of control after weeks of fighting them. In fact, one out-of-control blaze of 83,000 acres was the second largest forest fire in the nation. The largest fire, 94,000 acres, was burning in Utah. In addition, there were fires burning in Colorado, California, Nevada, and Washington. As of July 21, 2002, fires had blackened 3.5 million acres in the United States, about the total acreage burned in all of 2001, according to folks at the Interagency Fire Center. And there simply were not enough firefighters to go around.[123]

With the preceding in mind, it is imperative when in making decisions about patterns across the landscape that we consider the consequences of these decisions in terms of the generations of the future. Although the current trend toward homogenizing the landscape may make sense with respect to maximizing short-term profits, it bodes ill for the long-term, biological sustainability and adaptability of the land.

Here, it must be noted that the economic and ecological systems are perceived to operate on different scales of time, meaning that the long-term, detrimental effects to the environment caused by decisions made in favor of short-term profits are ignored. For this reason, it is important to remember, when considering diversity and its stabilizing influence, that it's the relationship of pattern, rather than numbers, that confers stability on ecosystems.[124] To create a sustainable, culturally oriented system, even a very diverse one, it is necessary to account for these co-evolved relationships, or the system has about as much chance of succeeding as has a sentence made out of randomly selected words.[125] This means a forest ecosystem must include ecological diversity as part of its long-term caretaking when it is converted to a culturally oriented ecosystem.

Ecological Diversity

Every ecosystem adapts in some way, with or without the human hand. The challenge is that our heavy-handedness precludes our ability to guess, much less know, what kind of adaptations will emerge. With such the case, it behooves us to pay particular attention to the "nuts and bolts" of diversity as both creator and protector of ecological back-up systems.

Biological diversity: A critical part of caretaking the public forests is to maintain and/or restore quality habitat based on the connectivity of habitat patches through maintenance and/or restoration of both latitudinal and elevational corridors for the seasonal migration of species, as well as the dispersal of young animals. Because it is important to know what species one is dealing with and what their habitat requirements are, each District must conduct periodic "presence/absence" surveys of both aquatic and terrestrial species under the direction of the staff fisheries biologist, the staff wildlife biologist, and the staff botanist. If necessary, either or both biologists and the botanist could have their portion of the survey conducted via contract. If, in addition to these standard surveys, a special one was needed that required the expertise of a herpetologist, entomologist, geomorphologist, mycologist, or some other discipline, it could be contracted out.

Genetic diversity: The protection of genetic diversity would be accomplished, as much as possible, through natural seeding in order to maintain and/or enhance the in-place gene pool and thus a forest's potential adaptability to environmental change, like that of climate. Because global climate change, such as warming, may force a forest to migrate upward in elevation, it is critical to maintain the necessary genetic steppingstones along potential routes of migration.

Functional diversity: Functional diversity is an outcome of species composition and the physical structures such composition creates and maintains in space through time. Functional diversity would be generally determined by the aggregation of treatments created across the landscape as caretakers emulate a mosaic of fire-created vegetational patterns.

Disease and Pests: Having discussed the basic components of ecological diversity, it is important to understand the role of disease and animal pests (both invertebrate and vertebrate) in the creation, modification, and maintenance of ecological diversity. I bring this up because disease and animal pests are often seen as doing little more than damaging live trees that then die and decompose— "stealing" short-term economic gain from those who would profit from the trees if they lived long enough and were in good enough physical condition to be cut and taken to a mill.

The flip-side is that, in the long term, every living system is to some extent self-nurturing by reinvesting biological capital in itself. By this I mean that part of every living system is always in the process of dying even as other parts are in the process of living. Within this cyclical nature of flow and ebb, give and take, life and death, each system tends to retain and recycle the elements of its dying and dead components to nurture those of its living and growing components. In this sense, there often is more life in a large, fallen, decomposing tree than there was in the live tree.

Another point about the importance of diseases and pests is the very fact that they sicken and kill trees. Declining and dead trees account for much of the ongoing diversity of ecological processes within a forest, processes that are largely hidden from sight and continually changing; processes that often require decades or centuries to complete, such as the addition of vital organic material to the soil.[126] In turn, declining and dead trees, both standing as snags and fallen, are important habitat over time for a changing clientele of plants, as well as invertebrate and vertebrate animals.[127]

Further, declining and dead trees open gaps in the canopy that let more light into the forest and stimulate a diversity of light-loving plants to grow that, in turn, add to the habitat diversity of a particular place in time and space. Fallen trees also add vertical relief to the forest floor that arrests soil from creeping down slope with the pull of gravity and aids the infiltration of water into the soil, where it is purified and stored over long periods of time.[128]

Although I could go on *ad nauseam*, I think these few points suffice to make it clear that diseases and so-call "pest species" have a variety of critical roles to play in maintaining the health of forests and so must be accommodated in how we treat forests as biological living trusts.

Ecological Back-ups: Ecological back-ups are the purveyor of system resilience. In turn, biological diversity, genetic diversity, and functional diversity, as well as the diseases and pests that help keep the forests in the cyclical motion of living and dying, are the bedrock of such back-ups. Hence, if diversity (biological, genetic, and functional) is accounted for in the patterns of vegetation and habitat across the landscape, ecological back-up systems will take care of themselves.

Natural-Area System: Beyond the ecological diversity discussed above, there are within every landscape special features that add to the overall diversity of an ecosystem, such things as caves, cliffs, talus slopes, and edaphic (= soil) habitats with endemic plants (those limited to small areas in geographical distribution), to name a few. Because these areas have a singular distinctiveness within the landscape, they require special care and must be maintained within a system of protected "natural areas."

In addition to these special features, which are fixed in location, there is a need to protect "legacy forests"—old-growth forests—as an unconditional gift of potential knowledge for the future. Other types of "legacy plant communities" are also necessary against which to measure the effectiveness of our human ability to emulate the patterns created by fire and the processes of various successional stages as professional caretakers discharge their duties in perpetuating the public's biological living trust.

Unlike the above-mentioned special features, legacy forests and legacy plant community types are free to move across the landscape in the course of time, which means the adequacy of their representation in space and time would be a major component of each generation's trusteeship duties. And one of the physical dynamics that keeps legacy forests and legacy plant community types on the move over time is fire.

Part of the process of performing the duties of trustees in caretaking public forests as a biological living trust is setting aside an ecologically adequate system of natural areas, both fixed features and "free-floating" plant community types. In so doing, all generations have repositories of species that, more often than not, are region-specific and processes that, more often than not, are worldwide in principle and application.

From such repositories, in addition to monitoring human-caused changes and maintaining habitat for particular species, it is possible to learn how to maintain and/or restore biological processes in various portions of the ecosystem. That is, reserves of indigenous forests, including old-growth forests, are the parts catalog and maintenance manual for forests—present *and* future.

Old-Growth Forests

There are many valid reasons for saving old-growth forests from extinction, as many perhaps as there are for saving tropical forests. One reason is that such forests as those of the Pacific Northwest are beautiful and unique in the world. Another is that such old-growth trees inspire spiritual renewal in many people and are among the rapidly dwindling, living monarchs of the world's forests. Old-growth forests are the oldest and largest living beings on Earth, and, as such, form a tangible link with the past and provide a spiritual ground in the present. Consider that trees can and do live for centuries or millennia; bristlecone pine, for example, is known to live more than 5000 years.

Of these bristlecones, author Jane Braxton Little writes: "Twisted with time, unspoiled and austere, the planet's oldest living inhabitants offer insights to environmental conditions back to the last ice age. Survivors of inconceivable eons and all-but-uninhabitable surroundings, they provoke a wonder about our time and place and purpose."

What would happen to our sense of continuity, our spiritual ground if all the remaining commercially available, old-growth forests were liquidated for short-term profits? Without the stability of their longevity, how would you cope with our dynamic, unpredictable Universe, the uncertainty of life? How would you feel if these centuries-old living beings were suddenly replaced with trees that were not allowed to grow much older than you? How would you feel if you knew that the oldest living beings on Earth had all been converted into money and you would never again see them, and your children and grandchildren would never have a chance to see them?

A third reason for saving old-growth forests is that they are unique, irreplaceable, and finite in number. They exist precisely once in the world because whatever is created in the future will be different—and centuries away, if ever. Large trees can perhaps be grown over two or three centuries, but they will be humanity's trees in society's cultural landscape—not Nature's trees in Nature's landscape. Although they may be just as beautiful as those planted and grown by Nature, they will be different in the human mind. They will not have the same mystique. And even if we start growing them today, neither we nor our children, nor our children's children for several generations will be here to see them—even if they were allowed to reach old age, which is doubtful.

A fourth reason for maintaining old-growth forests is that a number of organisms (e.g., the red-cockaded woodpecker, goshawk, northern spotted owl, northern flying squirrel, red tree mouse, and marten) find their optimum habitat in them; require the structures provided by the live old trees (e.g., the large declining trees, the large standing dead trees, the large fallen trees), or use them as refugia as in the case of the red tree mouse.[129] In other words, old-growth forests are tremendous storehouses of biological diversity. As such, they also are genetic reservoirs that harbor plants of potential use in medicine, agriculture, and industry.

A fifth reason to maintain old-growth forests is that such forests are the only living laboratories wherein we and future generations may be able to learn how to create sustainable forests, something no one has so far accomplished. Let's examine this latter reason in more detail.

As a living laboratory, old-growth forests serve many vital functions, including these four:

First, old-growth forests are indigenous and so form our link to the past, to the historical forest. As stated by South African General Jan Christian Smuts, "The whole, if one may say so, takes long views, both into the future and into the past; and mere considerations of present utility do not weigh very heavily with it."[130] Because the whole forest cannot be seen without taking long views both into the past and into the future, to lose the remaining old-growth forests is to cast ourselves adrift in a sea of almost total uncertainty with respect to the biological sustainability of the forests of the future. We must remember that knowledge is only in the past tense, learning is only in the present tense, and prediction is only in the future tense. To have sustainable forests, we need to be able to know, learn, and predict. Without old-growth forests, we limit our knowledge, eliminate many possibilities for learning, and greatly diminish our ability to predict.

Second, since we did not design the forest, we do not have a blueprint, parts catalog, or maintenance manual to help us understand and repair it. Nor do we have a service department, where the necessary repairs can be made. We cannot, under these circumstances, afford to liquidate the remaining old-growth forests that collectively act as a blueprint, parts catalog, maintenance manual, and service station. They are our only hope of understanding the long-term sustainability of our public forests.

Third, we are, after all these years of tinkering with the genetic "improvements" to trees, still playing "genetic roulette" with the forests of the future. What if our genetic engineering, cloning, streamlining, and simplifications run amuck, as they so often have around the world?

In spite of that possibility, most supporters of "genetic modification" downplay the difference between genetically engineered organisms and the techniques of time-honored selective breeding. They claim, for example, that the only difference between genetically engineered plants and traditional crops is that genetic manipulation is more precise, faster, and cheaper. Although this claim would seem to be good news, it is ecologically misleading and genetically irresponsible.

"Experiments have shown," writes Ricardo Steinbrecher, a genetic scientist and member of the British Society for Allergy, Environmental, and Nutritional Medicine, "that a gene is not an independent entity as was originally thought." Genetic engineers increasingly want to transform plants and animals from organisms with inherent novelty into "designed commodities," while ignoring the many unknown hazards.

Steinbrecher cites the example of a 1990 experiment in Germany, where the gene for the color red was taken from corn and transferred, together with a gene for antibiotic resistance, into the flowers of white petunias. The only expectation was a field of 20,000 red-flowering petunias. The genetically engineered petunias did indeed turn red, but also had more leaves and shoots, a higher resistance to fungi, and lower fertility. These unexpected results were completely unrelated to the genes for color and antibiotic resistance. Such unrelated, unexpected results have been termed "pleiotropic effects" that, by their very nature, are totally unpredictable.[131]

In this case, the pleiotropic effects were both clearly visible and easily identified without molecular analysis. But what happens if pleiotropic effects are not so obvious, if they unexpectedly affect the composition of proteins, the expression of hormones, or the concentration of nutrients, toxins, or allergens? Who is going to monitor all the possible pleiotropic effects before a genetically engineered plant is introduced into the environment or placed on our dinner plates? There are neither regulations nor voluntary guidelines and practices in place to check for pleiotropic effects, which makes cross-species cloning—including between humans and cattle, rabbits, and other species—particularly arrogant.[132] And even if there were regulations governing pleiotropic effects, how would one know what to look for?

And then there are the escapees. DNA from genetically engineered corn has shown up in samples of indigenous corn in four fields in the Sierra Norte de Oaxaca in southern Mexico. This finding is "particularly striking" say Ignacio Chapela and David Quist, researchers from the University of California, Berkeley, because Mexico has had a moratorium on genetically engineered corn since 1998.

The fact is that corn, genetically engineered to resist herbicides or to produce their own insecticides, threatens to reduce the variety of plants in that region of Mexico because they may be able to out-compete the indigenous species. "The probability is high," says Chapela, "that diversity is going to be crowded out by these genetic bullies." This type of unwanted genetic transference is termed "genetic pollution." In addition, the herbicide resistance could jump into weedy relatives and create "super weeds" that are beyond control, and plants that have been genetically engineered to produce their own insecticide can have serious, deleterious effects on insects and microbes in the soil that would also affect indigenous plants.[133]

For these reasons, indigenous forests, whether old or young, are imperative because only they contain the entire, unadulterated genetic code for living, healthy, adaptable forests.

Fourth, intact segments of the old-growth forest will allow us to learn what we need to know and make the necessary adjustments in both our thinking and our subsequent course of action to help assure the sustainability of our public forests. If we choose not to deal with the heart of the old-growth issue, which is the biological sustainability of present and future public forests, we will find that reality is more subtle than our understanding of it and that "good intentions" will likely give bad results—witness the suppression of fire.

Although there are many valid reasons to save old-growth forests, there is only one apparent reason for liquidating them: short-term economics. Economics, however, is the common language of Western industrialized society; so it is wise to carefully consider whether saving substantial amounts of well-distributed, old-growth forests on public lands is a necessary part of the equation for maintaining solvent *forest-dependent* industries, as opposed to just the timber industry.

The timber industry, as it's usually thought of, goes from the forest to the mill, but the United States—in fact the world as a whole—is founded largely on an interdependent suite of forest-dependent industries that individually and collectively rely much more heavily on abundant, clean water than they do on the growing and harvesting of wood fiber.

A forest-dependent industry is any industry that uses raw materials from the forest, including amenities and services like oxygen, water, electricity, and recreation, as well as commodities like migratory animals, such as salmon and steelhead. A forest-dependent industry also includes any industry that uses extractive goods like minerals, wood fiber, forage for livestock, resident fish and game animals, and pelts from fur-bearing mammals.

Some forest-dependent industries are based on amenities and services that are not extractive in the sense that the products either enter and/or leave the forest under their own volition. Such industries include the sport and commercial fisher who catches migratory salmon and steelhead in the ocean and rivers outside of the forest, the farmer who uses water to irrigate crops, the person who markets those crops, the electrical company that uses water converted to electricity, and the municipal water company itself.

Other forest-dependent industries are based on extractive products that are physically removed as raw materials from the forest and made available for refinement. Such industries include timber companies that cut trees; people who gather mushrooms commercially;[134] ranchers who graze livestock in forested allotments; miners who extract ore; hunters, fishers, and trappers who kill and remove forest-dependent wildlife.

Forest-dependent industries that refine the extracted products include carpenters, boatbuilders, artisan woodworkers, anyone who uses paper, meat cutters and packers, and furriers. Finally, these forest-dependent industries are all interwoven because each industry uses one or more of the other's products, such as water, electricity, wood fiber, red meat, and vegetables.

Considering the critical importance of forest-dependent industries to our social welfare, can we really afford to liquidate our remaining old-growth forests on public lands? It has often been said that we cannot afford to save old-growth forests because they are too valuable as a commodity and too many jobs are at stake in the timber industry. We must be careful here because scarcity not only increases the economic value of the remaining old-growth forests but also in-creases their ecological value. In the face of this scarcity, cutting the remaining old-growth forests will serve only a small proportion of the immediate genera-tion of adult humans involved in the timber industry, whereas protecting them will serve all individuals within all generations in the forest-dependent indus-tries. We must therefore be exceedingly cautious lest economic judgment—couched in political language about the urgent need to control the so-called "wildfires" brought about by decades of suppression—becomes the justification for liquidat-ing as many old-growth trees on public lands as possible in the name of "fire prevention."

It would cost $2.7 billion just to thin the 1.6 million acres of public forest in the Klamath Mountains of southwestern Oregon, not counting wilderness, roadless areas, and stands that are minimally stocked. If the old-growth trees are the primary ones to be cut, they would undoubtedly help pay for the thinning. As well, they would be an economic boon to proponents of the timber industry who have been frustrated for twelve years by court orders directing the Forest Service to follow the *intent* of the National Forest Management Act and protect habitat for fish and wildlife. Moreover, cutting the old-growth trees would *not reduce the risk of fire*.

On the other hand, while thinning understory shrubs and trees less than 21 inches in diameter at breast height (= the height of an average person's chest from the ground) would reduce the risk of fire, it would also cost an average of $1,685 per acre; whereas it costs $785 per acre for firefighters to control a 2,800-acre fire in southwestern Oregon. Yet one thing is clear, nothing that is done will stop forest fires, but it may reduce their intensity and to some extent control the loca-tions in which they burn.[135]

Although linear-minded folks in the timber industry cited environmental-ists in court for holding up the logging of national forests and thereby placed significant blame on them for the forest fires of 2002, a report by the General Accounting Office, the investigative arm of Congress, came to a different conclu-sion. The report (*Forest Service: Appeals and Litigation of Fuel Reduction Projects*, August 31, 2001), based on the examination of 1,671 treatments aimed at reduc-ing the level of fuels, said that a total of less than one percent of the government's attempts to thin national forests were challenged by environmentalists and other outside groups.[136, 137] The report made by the General Accounting Office is vastly different than that made by the U.S. Forest Service.

The Forest Trust, based in Santa Fe, New Mexico, compared the reports made by the General Accounting Office and the U.S. Forest Service and sent the results to Congress in 2002. The Forest Trust found the following biases in the data used by the Forest Service in its report (*Factors Affecting Timely Mechanical Fuel Treatment Decisions*, released to the media on July 10, 2002):

1. The Forest Service biased its sample by selecting only those projects that tended to be most frequently challenged. Specifically, the sample included mechanical treatments, but excluded all other forms of fuel reduction, such as prescribed fire.

2. The assertion on the part of the Forest Service that the reduction of hazardous fuel loads by mechanical means reflected an overall effort to reduce fuel loads is inaccurate. In 2001, mechanical treatments occurred only on fifteen percent of all the land treated by the Forest Service to reduce the loading of hazardous fuels, but the report analyzed only these mechanical treatments.

3. Of the 155 appealed projects the Forest Service said were directed toward the reduction of fuel loads, many were not, in fact, connected to the reduction of fuels, which compromised the integrity of the data.

4. Another error in the report is the inclusion of at least 37 projects to reduce fuel loads that the Forest Service did not make the General Accounting Office aware of.

5. The Forest Service did not use a consistent definition of "mechanical treatments" to reduce hazardous fuel loads; that is, 88 percent of the appealed projects listed in its report included commercial sales of timber. Further, the Forest Service neither explained how the 116 commercial timber sales contributed to the reduction of hazardous fuel loads nor why projects that more clearly did contribute to the reduction of hazardous fuels were not included.

6. The General Accounting Office examined 762 planned projects on 4.7 million acres of national forest land. Four hundred fifty-seven (sixty percent) of those projects had no environmental studies that could be appealed. One hundred eighty of the remaining 305 projects were challenged, but most were resolved within the ninety-day period required by the Forest Service. Thirty-nine took longer due to a shortage of staff personnel, an existing backlog of appeals, or to give contestants more time to negotiate a settlement. Of the 180 appealed projects, 133 proceeded unchanged, sixteen were modified, nineteen were blocked, twelve were withdrawn by the Forest Service, and a mere 23 (three percent of the total) were challenged in court.

7. Finally, according to the General Accounting Office, the Chief of the Forest Service can proceed with a treatment, such as thinning, despite appeals, if the project is deemed an emergency.

Nevertheless, there is still a call on the part of linear-minded thinkers for a "balance" in forest policy.[138] I have heard all my career about this "balance" from people with vested interests in liquidating the old-growth forest, and the "balance" always comes down to one thing—cutting as many old-growth trees as industry can get as fast as possible.

If industry is allowed to liquidate the remaining commercially available, old-growth forests, for whatever economic reason, the timber industry will be the baby thrown out with the bath water, and we will have further impoverished the quality of our public forests for future generations through the myopic drive for short-term profits. I advance this observation because some timber companies have been reported to cut down the largest, most commercially valuable trees in a number of fire-damaged forests, instead of the smaller ones that constitute the fuels most likely to burn and carry a fire.[139]

Unfortunately, this behavior on the part of the timber industry seems true to form. The last time the federal government suspended laws in the name of forest health was in 1995, when the Clinton administration approved a Republican-derived timber salvage program after large fires. Although the law passed Congress, it predictably led to the cutting of the *most commercially valuable, live, old-growth trees* in the Pacific Northwest.[140]

If such behavior is not acceptable, how should we caretake old-growth forests? The first step in treating an old-growth forest is to characterize it in order to know what its defining ecological features are; an example of such characterization can be found in: "Ecological Characteristics of Old-Growth Douglas fir Forests."[141] Once the old-growth characteristics of species composition, structures, and processes of a given forest have been identified, they would form the underpinnings of the treatment plan.

To deal with an old-growth forest as an interactive whole, each District office would have the pertinent literature in its library, such as the book titled: "Wildlife-Habitat Relationships in Oregon and Washington."[142] This book has an incredible wealth of integrated data that can augment one's understanding of the various types of forests on public lands in both states, as well as interactive components, such as fish and wildlife, and other information that can help caretakers treat old-growth forests as irreplaceable components of a biological living trust that includes roadless areas.

Laws, Policies, and Roadless Areas

The methods we choose for "caretaking" public lands are based on and controlled by laws and policies, both stated and unstated. Each policy is either a true or a false reflection of public law; in that sense, the methods of caretaking public lands may be more or less cooperative and environmentally benign or more or less competitive and environmentally malignant. Whichever methods we choose, we do so consciously. To illustrate, while I was still working for the Bureau of Land Management, the controversy over protecting the northern spotted owl arose. Although the owl became a protected species under the Endangered Species Act, which meant that its habitat of old-growth forests was also to be protected, the state director of the Bureau at that time had an unwritten policy that *all* old-growth forests on Bureau lands in western Oregon and Washington were to be clear-cut within thirty years—a policy he pursued until he retired, over the constant objections of the Bureau's wildlife biologists. Clearly, policy can be a seriously weak link within agencies.

This being the case, a Synthesis Team, as described earlier, would be assembled to examine all laws—and policies—applicable to public lands. The Team would include systems-thinking lawyers; ecologists; economists; sociologists; residents from rural areas, suburbia, and the inner city; and Indigenous Americans. Once assembled, it would evaluate the laws and policies to see which ones actually protect public lands, which are merely conflicting, and which are outdated.

The Team's charge would be to sort out those laws that actually protect public lands, those that do not, and to see which of the laws take precedence. With this background, the Team would produce two reports. In one, the Team would build a strategy for the immediate caretaking of public lands with the best possible set of applicable laws and policies that are currently aligned with the land's ecological protection. Part of this report would be a review of the strengths and weaknesses of each law or policy as it pertains to the legal aspects of caretaking public lands as a biological living trust. Such an analysis would allow the previously mentioned "weakest-link theory" to be used as a means of continually improving the trusteeship of our public lands by crafting appropriate policies as guides for thought, decision, and action.

In the second report, the Team would bring to the fore all laws and policies that conflict with that protection and would highlight those laws and policies that must be updated or deleted from the books because they contradict the intent of a biological living trust, such as the 1872 mining law. The Team would then work with legislators and the public to clarify and update the legal aspects of caretaking public lands as a biological living trust.

The upshot: If we are to bequeath our children and the generations of the future a chance for a quality life, we must strive toward a vision that restores the health and adaptability of the landscape in the present for *all* generations. Such a bequest means creating policy that, in fact, reflects the law and its mandates in letter *and* spirit.

Until the laws and policies have been studied and revised as needed, a moratorium needs to be placed on the disturbance of roadless areas. A roadless area is a forest that is unroaded or lightly roaded, with no evidence of previous logging, and of sufficient size (500 acres or larger) and configuration to maintain ecological integrity. This action is necessary because, in our burgeoning, product-oriented society, one of the most insidious dangers to indigenous forests—those with minimal, disruptive human intrusion—is the sadly mistaken perception that there is no value in maintaining such a forest for its intrinsic potential. By intrinsic potential, I mean its value as a blueprint of what a sustainable forest is, how it functions, its educational value, its spiritual value, or any other value that does not turn an immediate, visible, economic profit.

Although some roadless areas have survived in the modern world, they are not only constantly changing but also in danger of natural, catastrophic disturbance, such as large, intense fires. Consequently, they cannot simply be left alone because, should one burn, it is gone forever. True, the acreage is still in place and the land is still roadless, and the trees will eventually grow back, but the indigenous forest is gone—and that is the point.

Furthermore, roadless areas are at the most natural end of the continuum between naturalness and culturalness. As such, they are critical areas in which to learn about the long-term processes and trends of a naturally developing forest. It is thus imperative to emulate the disturbance regimes to keep them healthy so they can develop along their ordained trajectories and protect them, as much as possible, from catastrophic events so future generations can benefit from their existence. In other words, these roadless areas are far too important to lose.[143]

Because of their great importance, roadless areas would be treated in the planning phase the same as old-growth forests, with the added criterion that *they would remain roadless*. Accordingly, any treatment to "fire-proof" them, such as thinning understory fuel ladders and clearing vegetation that could easily ignite in a dry-lightning storm and carry a fire, would be done with horses and, if necessary, with the additional aid of helicopters. With imagination, existing knowledge, new information, and a shared vision of what a given forest as a biological living trust can be, I think a combination of the available and developing silvicultural techniques can be used to achieve the desired outcome—*with the exception of clear-cutting*.

There would be *no clear-cutting* or *high-grading* of public forests under any circumstances because clear-cutting emulates nothing in Nature. Clear-cutting is purely and simply an economic device to reap the greatest income in the most expedient manner.

With a template of landscape-scale fire patterns, or another of Nature's disturbance regimes, to guide the caretaking of public forests, I think these silvicultural techniques can be imaginatively used—with ecologically sound justification—to emulate a given fire-designed pattern: uneven-age treatments, individual-tree selection, small seed-tree cut, group selection, variable retention/variable density thinning, shelterwood cut, and the leaving of an adequate legacy of snags and large woody debris to meet the soil's requirement for the reinvestment of biological capital, as well as *prescribed fire*, which brings us to "riparian areas."

Riparian Areas

Because a sustainable supply of water is increasingly critical to humans, fish, and wildlife, due to our burgeoning human population and the growing uncertainty brought about by the changing global climate, protecting every body of water on public lands for its capacity to store this precious liquid is crucial. To this end, a permanent moratorium must be placed on the construction of roads near bodies of water. Over time, as many roads as possible that parallel a stream or river must be closed. When feasible, other roads must be relocated to the tops of ridges, even if it increases the length of a given road. All culverts and bridges rendered unnecessary must be removed and the roads obliterated and permanently abandoned.

With respect to the moratorium on logging within riparian areas, that must be reviewed on a case-by-case basis. Where treatment is necessary to maintain a functional riparian area or to restore a degraded one, such treatment must be designed and carried out within sound constraints of Nature's ecological principles.

Riparian areas (also referred to as "riparian zones") can be identified by the presence of vegetation that requires free or unbound water and conditions more moist than normal in summer. These areas may vary considerably in size and the complexity of their vegetative cover because of the many interactions that occur between and among the source of water and the physical characteristics of the site. Such characteristics include gradient, aspect of slope, topography, soil, type of stream bottom, quantity and quality of the water, elevation, and the type of plant community.

Riparian areas share the following common denominators: (1) they create well-defined habitats within much drier surrounding areas, (2) they make up a minor portion of the overall area, (3) they are generally more productive than the remainder of the area in terms of the biomass of plants and animals, (4) wildlife use riparian areas disproportionately more than any other type of habitat, and (5) they are a critical source of diversity within an ecosystem.

There are many reasons why riparian areas are so important to wildlife, but not all can be attributed to every area. Each combination of the source of water and the attributes of the site must be considered separately:

1. The presence of water lends importance to the area because habitat for wildlife is composed of food, cover, *water*, and space. Riparian areas offer one of these critical components, and often all four.

2. The greater availability of water to plants, frequently in combination with deeper soils, increases the production of plant biomass and provides a suitable site for plants that are limited elsewhere by inadequate water. The combination of these factors leads to increased diversity in the species of plants and in the structural and functional diversity of the biotic community.

3. The dramatic contrast between the complex of plants in the riparian area with that of the general surrounding vegetation of the upland forest or grassland adds to the structural diversity of the area. For example, the bank of a stream lined with deciduous shrubs and trees provides an edge of stark contrast when surrounded by coniferous forest or grassland. Moreover, a riparian area dominated by deciduous vegetation provides one kind of habitat in the summer when in full leaf and another in the winter following leaf fall.

4. The shape and size of many riparian areas, particularly the sinuous nature of streams and rivers, maximizes the development of a quality edge effect that is so productive in terms of wildlife.

5. Riparian areas, especially those in coniferous forests, frequently produce more complex edges within a small area than would otherwise be expected based solely on the structure of the plant communities. In addition, many strata of vegetation are exposed simultaneously in stair-step

fashion. This stair-stepping of vegetation of contrasting forms (deciduous versus coniferous, or otherwise evergreen shrubs and trees) provides diverse opportunities for feeding and nesting, especially for birds and bats.

6. The microclimate in riparian areas is different from that of the surrounding area due to increased humidity, a higher rate of transpiration (loss of water) from the vegetation, more shade, and increased movement in the air. Some species of animals—including humans—are particularly attracted to this microclimate.

7. Riparian areas along intermittent and permanent streams and rivers provide routes of migration for wildlife, such as birds, bats, deer, and elk. Deer and elk frequently use these areas as corridors of travel between high-elevation summer ranges and low-elevation winter ranges.

8. Riparian areas, particularly along streams and rivers, may serve as forested connectors between and among habitats, such as high- or low-elevation grasslands that may border a mid-elevation forest. Wildlife may use such riparian areas for cover while traveling across otherwise open areas. Some species, especially birds and small mammals, may use such routes in dispersal from their natal habitats due to pressures caused by overpopulation and/or shortages of food, cover, or water. Riparian areas provide cover and often provide food and water during such movements.[144]

In addition, riparian areas supply organic material in the form of leaves and twigs that become an important component of the aquatic food web. Riparian areas also supply large woody debris in the form of fallen trees that form a critical part of the land-water interface, the stability of banks along streams and rivers, and instream habitat for a complex of aquatic plants, as well as aquatic invertebrate and vertebrate organisms.

Setting aside riparian areas as undeveloped open space means saving the most diverse, and often the most heavily used, habitat for wildlife in proximity to a human community. Riparian areas are also an important source of large woody debris for the stream or river whose banks they protect from erosion. Furthermore, riparian areas are periodically flooded in winter, which, along with floodplains, is how a stream or river dissipates part of its energy. It is important that streams and rivers be allowed to dissipate their energy; otherwise, floodwaters would cause considerably more damage than they already do in areas of human habitation.

A floodplain is an area, usually of boulders, gravel, or sand and largely devoid of vegetation that borders a stream or river and is subject to flooding. Like riparian areas, floodplains are critical to maintain as open areas because, as the name implies, they frequently flood. These are areas where storm-swollen streams and rivers spread out, decentralizing the velocity of their flow by encountering friction caused by the increased surface area of their temporary bottoms, both of which dissipate much of the floodwater's energy.

It is wise to include floodplains within the matrix of riparian areas for several other reasons: (1) they will inevitably flood, a circumstance that puts any human development at risk, regardless of efforts to steal the floodplain from the stream or river for human use (witness the Mississippi River); (2) they are critical winter habitat for fish; (3) they form important habitat in spring, summer, and autumn for a number of invertebrate and vertebrate wildlife that frequent the water's edge; (4) they can have important recreational value, and (5) they trap sediment and thereby help to keep streams and rivers flowing clean and clear.[145]

Ecological Restoration

The key to and the value of ecological restoration are the thought and act of putting something back to a prior position, place, or condition. That much is clear enough. But why should we humans bother putting something back toward the way it was? Why try to go backward in time when society's push is forward, always forward. The answer draws on two paradoxes: backwards is sometimes forward and slower is sometimes faster.

In our drive to maximize the harvest of Nature's bounty, we, in the United States, strive only for a *sustained* or *ever-increasing* yield of products, and we are intensively altering more and more acres worldwide to that end. We cannot have a sustainable yield of anything, however, until we first have a sustainable ecosystem, such as a forest, to produce the yield. In practice, we tend to think it a tragic economic waste if Nature's products, such as wood fiber or forage for livestock, are not somehow used by humans but are allowed instead to recycle in the ecosystem, compost as it were. And because of our paranoia over "lost profits" — defined as economic waste — we extract far more from the ecosystems than we replace. We will, for example, invest capital in another crop, but not in maintaining the health of the ecosystem that produces the crop. This scenario is in the tradition of our Western industrialized culture, and, through it, much of the lithosphere, biosphere, and atmosphere are being degraded.

This brings me directly to the value of restoration as a means of changing the way we think and as a means of changing the way we relate to the ecosystem we inhabit, whichever one it is. Basically, restoration helps us to understand how a given ecosystem, such as a forest, functions: as we put it back together, as we go backward in time to reconstruct what was, we learn how to sustain the system's ecological processes and its ability to produce the products we desire, now and in the future.

Similarly, restoration helps us to understand the limitations of a given portion of the ecosystem. As we put it back together, as we slow down and take time to reconstruct what was, we learn how fast we can push the system to produce products on a sustainable basis without impairing its ability to function.

Thus, the very process of restoring the land to health is the process through which we become attuned with Nature and, through Nature, with ourselves. Restoration, therefore, is both the means and the end, for as we learn how to restore the land, we heal the ecosystem, and as we heal the ecosystem, we heal the deep geography of ourselves. At the same time, we also restore both our options for products and amenities from the land and the options for future generations. This act is crucial because our moral obligation as human beings is to maintain options for future generations. To this end, maintenance and restoration are the heart and spirit of caretaking our public forests. I use the word "spirit" advisedly because it is derived from the Greek word for "breath," which denotes life.

We, as citizens, must learn to understand and accept that the "sustainability" of a forest, or any ecological system for that matter, is like a "free spirit," an ever-elusive prize that, like a horizon, continually retreats as we advance. This "dance of approach and retreat" causes me to think of "sustainability" as the duty of each generation to pass forward to the next as many positive options for safe keeping as humanly possible. This notion requires clarity of mind because it means that we, the adults, must finally come to grips with the fact that each generation must pay its own way—beginning with us, here, now. The cost of our presence on Earth must be accounted for in how we caretake the ecosystems that we and all generations to come must rely on for our survival. By this I mean that all debts incurred by the generation in charge must be paid by that generation—not passed forward as an ecological mortgage that encumbers the social-environmental welfare of the future.

To achieve the level of consciousness and the balance of energy necessary to maintain the sustainability of ecosystems, such as our public forests, we must focus our questions, both social and scientific, toward understanding the ecological governance of those systems and our place within that governance. Then, with humility, we must develop the moral courage and political will to direct our personal and collective energy toward living within the constraints defined by the ecological principles that govern ecosystems—not by our economic-political ambitions.

The systems we are designing—or redesigning—by our existence in, and our interaction with, our surroundings are continually changing the environment (all of it, if in no other way than through pollution of the air). Consequently, conditions prior to the arrival of Europeans in North America (of which we know little) are largely irrelevant because the compounding environmental influences of today's huge human population and its permanent developments have, in many ways, limited the possibilities of ecosystem restoration. Added to our current environmental dilemma, is the fact that indigenous populations were much smaller and more nomadic than the mega-populations of today. Moreover, the ecological systems with which we daily interact are becoming ever further removed from the types of ecological balances that characterized pre-European conditions.

Our challenge today is to mature sufficiently in personal and social consciousness to recognize a functionally healthy and sustainable ecosystem when we see it and then to maintain it as such. Beyond that, we need to restore functionally degraded ecosystems to the greatest sustainability we are capable of. The sustainability of which I speak is a process, a journey toward the ever-increasing consciousness that we humans must acquire in order to learn how to treat our environment for the benefit of all generations. Sustainability is *not* an absolute— *not* a materialistic endpoint.

Although sustainability is not a condition in which a compromise can be struck, the decisions leading toward sustainability often necessitate compromise. Seeking sustainability to a degree, which may appear to be an innocuous compromise, defeats sustainability altogether. Leave one process out of the equation or in some other way alter a necessary feedback loop, and the system as a whole will gradually be deflected toward an outcome other than the one originally intended.

Consider that humans have changed the dynamics of every landscape with which they have interacted and have been doing so for thousands of years. The history of England is an example.

About 330,000 years ago, long before people arrived in the Americas, the early human, *Homo erectus*, was already living in what today is England, which was at that time attached to the European continent. As the Pleistocene Epoch drew to a close between 12,000 and 10,000 years ago, the ice withdrew, though not in a single smooth recession. During this time, Paleolithic cultures of "modern humans" (old stone-age, earlier than 12,000 years ago) occupied the warmer places in the south of what today is England, where they seem to have had an ecological impact with their selective dependence on wild horses and reindeer for food and raw materials.

As the climate ameliorated, the trees that had survived the glaciation in southern Europe and the Caucuses gradually returned to the once-glaciated areas until climax-stage mixed deciduous forest was established about 8,000 years ago. Then, about 7,000 years ago, that part of the European continent that today is England separated from the mainland. Mesolithic (middle stone-age, between 12,000 and 5,000 years ago) hunter-gatherers still inhabited the newly formed island, and they remained until the coming of agriculture about 5,000 years ago.

The earliest cultural landscapes of the area—those purposely manipulated with fire—were formed in the middle to late Mesolithic period, between 7,000 and 5,000 years ago. As far as we know, that first cultural landscape came from the conversion of a mixed deciduous forest into a mosaic of high forests, open-canopy woodlands, and grassy clearings with fringes of scrub and bracken fern, patches of wet sedge, and bogs of peat. Among these habitats, groups of late Mesolithic peoples, without knowledge of crop-based agriculture, moved about gathering food.

The coming of agriculture from Asia, around 5,000 years ago, was one of the great turning points in Western Europe, for it was the beginning of the Neolithic culture (new stone-age, which began with the advent of agriculture). For Neolithic people, the advent of agriculture was not gradual. Instead, the full complement of agricultural tradition and myth probably came as a developed package from the East, even if accessory hunting persisted. The model of the earliest agriculture in Western Europe is a mosaic of small clearings that were abandoned as the fertility of the soil became exhausted or the weeds became too bothersome. As new clearings were made, abandoned ones reverted to forest and were again cleared only to be once more abandoned, and finally cleared with a sense of permanence.[146]

I bring this up because (with respect to caretaking public forests in the United States) there has been an increasing emphasis in recent years on the evolution of "natural ecosystems," as though only those that somehow evolved without a human influence would qualify. The idea of "natural" has been perpetuated by writers who have created the romantic myth of Indigenous Americans who somehow had the wisdom and self-control to live in perfect harmony with Nature, taking only the bare minimum of what they needed to survive and, by inference, voluntarily keeping their own populations in check. It has also been assumed that predators and their prey were in a perfect balance, that Nature's ecological disturbance regimes either did not exist or did not have any affect on the great American landscape until the Europeans invaded the continent—hence the idea of a "climax" ecosystem, one that is indefinitely stable. Clearly, such romantic ideas would render unacceptable any perceived human disturbance to the "balance of Nature."

With this background, the big question before those who engage in ecological restoration is: restoration to what? This is an important question because as we restore the land to health, we advance our sense of consciousness and thereby rediscover our inseparable connection to Nature. In the process, we will discover that cumulative circumstances have made it all but impossible to revert modern landscapes to those of old, which does not mean they cannot be healed. It only means that we must with wisdom choose the reason we restore them and to what condition.

Still, there is much insistence on the part of some people that ecological restoration of public forests should take them back to pre-European times, a proposal that is neither feasible nor possible for three reasons:

1. *We do not know what the conditions were prior to the European invasion that began with Christopher Columbus and the Spanish in 1492.* The first reason is self-evident. We have no way of knowing what the conditions were in North America north of Mexico prior to the landing of Columbus and the Spanish. Consequently, the desire of returning the ecological conditions of public forests to pre-European times is moot—there simply are no records.

2. *Whatever the conditions were, they reverted toward the "wild" side between the time the Spanish landed in 1513 in what is now Florida and when the British landed on the North American continent in the early 1600s. They reverted even more toward the "wild" side by the time Lewis and Clark made their historic trek across the North American continent (1803-1805), and still more by the time the Oregon Trail was in full use in the 1840s.* By 1492, indigenous peoples throughout the Americas had modified the extent and composition of the forests and created and expanded grasslands through the use of fire. In addition, indigenous peoples of the Americas rearranged micro-relief through countless human-created earthworks. Agricultural fields were common, as were houses, towns, roads, and trails.

In the Amazon River Basin, for example, the indigenous people demonstrated cooperation and coordination through sophisticated land-use planning, architecture, and engineering in transforming the jungle into a system of towns, villages, and smaller hamlets, all laced together with precisely designed and constructed roads, some more than fifty yards wide. Moreover, the roads went from one point to another in a straight line. All this "development" was done prior to 1492 in order to sustain a widespread culture based on farming.

Indigenous Americans changed their environment through a variety of means, some of which were so subtle that European settlers mistook the altered landscapes for ones untouched by human hands. They used such methods as: (1) creating raised and/or terraced fields for raising crops; (2) establishing complex systems of irrigation; (3) using fire to create park-like areas, where it was easy to hunt, collect nuts, and cultivate crops; (4) using fire to create and maintain grasslands for hunting big game animals; and (5) constructing major cities, such as those of the Aztecs, Mayans, Incas, and other peoples.

The size of indigenous populations, associated deforestation, and prolonged intensive agriculture led to severe degradation of the land's carrying capacity in some regions. Such was the case in central Mexico, where by 1519 the pressures to produce food may have brought the Aztec civilization to the verge of collapse even without the Spanish invasion. There is good evidence that severe erosion of the soil was already widespread and not solely the result of plowing the land, grazing it with livestock, and deforestation carried on by the Spanish. So it seems that the degradation of the land, in some areas at least, was as much a matter of long-term intensive use by indigenous peoples as it was intrusion by the Spanish.

All of these human activities had local effects on soil, microclimate, hydrology, and wildlife. Accordingly, the landscape of 1492 had all but vanished by 1750, not because the Europeans superimposed their design on it, but because the indigenous peoples had declined precipitously in numbers through the war waged on them by the invaders and through diseases the invaders brought with them to which the indigenous peoples had no immunity. Not surprisingly, the landscape of 1750 appeared to be more "pristine"—less humanized—than that of 1492.

Whoever watched as Columbus came ashore, if any indigenous people did, witnessed the beginning of a conquest destined to cause the greatest destruction of human lives in history. When the Spanish landed in the islands of the Caribbean in 1492, the indigenous population of the Americas was about 53.9 million, and is thought to be divided as follows: 3.8 million indigenous peoples in North America, 17.2 million in Mexico, 5.6 million in Central America, 3.0 million in the Caribbean, 15.7 million in the Andes, and 8.6 million in the lowlands of South America. As you consider these figures, bear in mind that the landscape of 1492 reflected both the extant population of the times and the cumulative effects of a growing population over a period of perhaps 40,000 years or more.

The decline of indigenous peoples, once it began, was rapid and severe—probably the single greatest demographic disaster in history. With European disease as the primary killer, populations of indigenous peoples fell by ninety percent or more in many regions, particularly the tropical lowlands, during the first century after the initial invasion.

The estimated declines of indigenous populations during this time are:

Hispaniola (formerly Haiti, an island in the West Indies) dropped from 1 million in 1492 to a few hundred within fifty years (more than ninety percent). In Peru, the population declined from nine million in 1520 to 670,000 in 1620 (92 percent), in the Basin of Mexico from 1.6 million in 1519 to 180,000 in 1607 (89 percent), and in North America from 3.8 million in 1492 to 1 million in 1800 (74 percent). Collectively, the indigenous population declined from 53.9 million in 1492 to 5.6 million in 1650, which amounts to an 89 percent reduction.

This decline is not surprising when you consider that in what today is the United States, the Spanish controlled the land in the mid-1500s from the Carolina coast as far north as La Charrette, the highest settlement on the Missouri River, to at least San Francisco Bay in California and thereby exposed the indigenous population to European diseases. Decimation of the indigenous population through Spanish-style conquest and the spread of European diseases affected the human-influenced landscape accordingly, although there was not always a direct relationship between the density of a human population and its impact.

These circumstances point to a significant environmental recovery of the land by 1750, with a commensurate reduction of indigenous cultural features. Some of these changes are evident in the historical accounts of travelers, such as Columbus who sailed along the north coast of Panama on his fourth voyage in 1502-03. During this voyage, his son, Ferdinand, described the land as well-peopled, full of houses, with many fields, open land, and few trees. Lionel Wafer, in contrast, found most of the Caribbean coast of Panama covered with forests and unpopulated in 1681. And so it was all over the Americas: forests grew back and filled in, soil erosion became stabilized, agricultural fields became occupied by scrubs and trees, and indigenous earthworks became overgrown.

By 1650, indigenous populations had been reduced by about ninety percent in the hemisphere, whereas the numbers of Europeans were not yet substantial in 1750, and European settlement had only just begun to expand. As a result, the fields of indigenous peoples were abandoned, their settlements vanished, forests recovered, savannas retreated as forests expanded, and the subsequent landscape did indeed appear to be a sparsely populated "wilderness."[147]

Here it is important to point out that, prior to the invasion of Europeans, human impact on the environment was not simply a process of increasing change in response to the linear growth of the indigenous populations. Instead, the landscape was given time to rest and recover as cultures collapsed, populations declined, wars occurred, and habitations were abandoned. The effects of human activities may be constructive, benign, or destructive, all of which are subjective concepts based on human values, but change is continual, albeit at various rates and in various directions. All changes are nevertheless cumulative. Even mild, slow change can show dramatic effects over the long term.

Although there was, of course, some localized European impact prior to 1750, after 1750, and especially after 1850, populations of European Americans expanded tremendously, exploiting the resources more intensively and greatly accelerating their modification of the environment. To exploit the land and the remaining indigenous peoples with "moral" impunity required a rationalization — and so was born the grand "American Myth."

The grand American myth in the United States is one of imagined pristine Nature across an entire continent of wilderness filled with wild beasts and savages, which was probably not as difficult for settlers to conquer as has been imaginatively conceived. The "ignoble savage," nomadic and barely human, was invented to justify stealing the land from the few remaining indigenous North Americans and to prove they had no part in transforming an untamed wilderness into a civilized continent. When the Europeans walked into a forest, which they described as "parklands," they did not see the indigenous peoples creating them through the use of fire, nor did they see the prairie-like conditions of large, open valleys, such as the Willamette Valley in Oregon or the savanna in Wisconsin being maintained by the indigenous peoples, also with the use of fire. In fact, the Indian use of fire may have been the most significant factor in designing the great American landscape first seen by the invaders.[148] They did not see what came before they arrived on the scene and put the best spin on what they saw by assuming it was "natural"—which meant, and still means to many people, untouched by the defiling hands of humans.

With the indigenous peoples branded as indolent and incapable of the art of civilization, the United States was built, according to official historical texts, by Puritan saint, yeoman, mountain man, frontiersman, pioneer, sodbuster, cowboy, and lumberjack, all of whom are painted as overachieving, self-glorifying Americans. Today, whether denigrated as subhuman "savages" or idealized as "Native Americans" living in perfect harmony with their environment, the indigenous peoples of North America are given no credit for having evolved well-developed cultures that molded the North American landscape into open woodlands in the east and stately conifer forests in the west, for creating and maintaining much of the continent's grasslands, and for transforming hardwoods in the southeastern United States into piney woods through the use of fire.

In essence, whether through ignorance or, perhaps and more probably, through deliberate "informed denial" of the role played by the indigenous peoples in shaping the landscape over millennia, the Europeans (such as the British) downplayed the evidence of cultivation and permanent settlements. To morally and legally justify stealing the land from the indigenous peoples, the British had to do two things: (1) portray the indigenous peoples as nomadic hunting-gathering savages that were forever an incorrigible and uncivilizable part of the wild and untamed continent and (2) demonize the indigenous peoples as indolent, untrustworthy, murdering heathens—and hence the enemy of pious civilized people like the British who, incidentally, paid bounties for the scalps of indigenous men, women, and children to encourage people to shoot them on sight. In other words, if the indigenous people were not *really* using the land, they had no title to it, and the British could feel justified in taking it.

In fairness, and paradoxically, it must be stated that the European Americans undoubtedly found much more "forest primeval" in 1850 than had existed in 1650. Nevertheless, writers and historians, both consciously and unconsciously, denied evide indigenous people's having created a cultural landscape over millennia in order to ennoble the European enterprise of pirating the land. This approach to creating a national myth for the fledgling United States—which supposedly is founded on the premise of human equality—helped European Americans envision themselves as single-handedly taming a vast, wild continent, to which they would ultimately bring civilization—"Manifest Destiny."

Despite how the European Americans envisioned themselves, Robert Hine, professor of history at the University of California at Riverside, wrote that ". . .it takes time to beat the Indian out of the soil, and how could that typical pioneer tame the land if he was always moving on, always settling, never settled! He seemed to love his land so little that he was willing to sell or leave at the drop of a hat. One observer of the West, a man from Scotland, where land was cherished, found Americans without qualms in abandoning their land."[149]

3. *Even if we had an idea of what the pre-European conditions were, we could not physically go back to them.* It is an inalienable truth that we can physically go back in space to a particular place, but we can *never go back in time* to who we were at a given moment in the past or what the particular circumstances were. Consequently, trying to caretake the public forests by ecologically restoring them to a condition we do not know in a time we cannot recapture is a physically impossible task.

In fact, this kind of ecological restoration is moot in terms of today's world for many reasons; a few are:

- population—there are far more people on the North American continent now than prior to the Spanish invasion
- pollution—today, in contrast with pre-European times, the entire North American continent is polluted, a condition that dramatically affects what can be done in the name of ecological restoration
- capitalism—a human invention foisted onto the North American continent by the European invaders and used to feed the insatiable human appetite for material goods even as the competition it spawns for raw materials destroys the ecosystems that sustain the economy
- rights of private property—the European concept of the absolute rights of private property, as opposed to the indigenous practice of shared rights to use communal land, does much to preclude the ecological restoration of a seamless forest
- technology—technological changes that have grossly and irreversibly altered the entire landscape of the United States

So the question is: *what* kind of restoration will benefit us today and the children of tomorrow and *why?* This is at once an intelligent question and a *wise* one because it is both present and future oriented. Suppose, for example, that a community has a forested water-catchment that has been degraded over many years by exploitive forestry and is no longer storing water late into the summer; as a result, the municipal reservoir holds progressively less water each year. And let's suppose further that a remnant population of bull trout is barely surviving in the streams of the water-catchment. Bull trout are highly *adapted*, unlike humans who are highly *adaptable*. While adaptable means having the *ability* to adapt to change, adapted is already a *fait accompli* with respect to given circumstances. Being highly *adapted* to clear, clean, cold water, bull trout are quite rigid in their habitat requirements.[150] So what, you might ask, does a tiny, remnant population of bull trout have to do with ecological restoration?

In this case, the demise of the bull trout is closely tied to the degradation of the water-catchment through the years of abusive forestry practices, such as repetitive clear-cutting, that have affected the quality of the trout's habitat, both directly and indirectly. A direct affect of clear-cutting has been the removal of trees right down to the edges of the streams, a practice that allowed the water to be warmed by the sun, something the bull trout cannot tolerate. Indirectly, the clear-cutting affected the trout's habitat by changing when and how the snows melt. In the past, snow melted slowly, and the water infiltrated into the soil, where it was purified, cooled, stored, and released into the streams all year, maintaining good late-summer flows to the benefit of both the bull trout and the people. But now the snow melts early and fast each year, causing the water to flow over the surface of the ground, collecting and transporting sediment into the streams as it goes. The accelerated melting degrades the purity of the water and decreases the late-summer flows because the water is largely gone by late June, the result being less water by July for both the bull trout and people. And less water means warmer water for the trout, an intolerable situation.

The economic effect of less water in late summer is devastating to the local forest-dependent industries, so the people decide to rectify the situation, but are not sure how. As it turns out, a fisheries biologist from the local university attends one of the meetings held to discuss the situation. When it comes his turn to speak, he explains the direct and indirect connections among the repetitive clear-cutting of the forest, the shortage of late-summer water, the degradation of the bull trout's habitat, and the fact that the bull trout is highly adapted to a narrow set of circumstances. He then goes on to explain that the community could restore its water-catchment by restoring the bull trout's habitat, which, in effect, is the barometer of the water-catchment's health. And so, with the guidance of the fisheries biologist, the restoration project begins—with a single purpose (restore the water-catchment to ecological health, the *what*), using the bull trout as the water-catchment's "canary in the coal mine" (a practical, measurable endpoint that can be monitored over time). To restore and maintain year-round supply of water for the sustainability of the forest-dependent industries is the *why*. This set of circumstances answers the compound question of restoration: the *what* and the *why*.

There is an interesting caveat to restoring areas of forest, both internally and externally. As a forest is degraded by such exploitive forestry practices as clear-cutting, the forest's internal microclimate is destroyed. As successive areas of adjacent forest are clear-cut, the region's local climate can be altered in a dramatic, and, in some cases, irreversible way.

By way of illustration, imagine the forests of the Cascade Mountains in western Oregon and Washington, the Klamath and Siskiyou Mountains in southwestern Oregon, the Coast Range in western Oregon and Washington, and the Olympic Mountains in northwestern Washington all burning in one year. According to Boone Kauffman, once a fire ecologist at Oregon State University, that is "roughly equivalent to the amount of land that burned in only one year in the Brazilian Amazon." Each year, an area that is eighty percent the size of the state of Oregon burns in the Brazilian Amazon alone.

Although the major cause of deforestation leading to such extensive burning is the conversion of tropical forests to pastures for cattle, simply cutting timber also causes problems because once the canopy is opened, the understory environment changes drastically and the forest can no longer sustain itself. Never in the history of humanity has so much of the world's tropical forests been disturbed in such a foreign and catastrophic way on such a large scale as during the last thirty-some years. The significance of this statement lies in the fact that tropical rain forests—one of the world's oldest ecosystems—occupy only seven percent of the Earth's surface but are home to more than fifty percent of all the Earth's known species. What does this mean in terms of the Amazonian tropical forest?

An intact rain forest creates its own internal and external climate in which about half of all the rainfall originates from moisture given off by the forest itself. When large areas are deforested, local and regional climatic patterns change. Once the forest is gone, drought is likely to occur, increasing the probability of fire, while decreasing the probability that the original type of forest will ever return.

The environment in the deforested areas of the Amazon has been altered to such an extent that the ecological processes that once maintained the tropical forest are unraveling in an irreversible change. In fact, the environmental conditions change swiftly and dramatically when the forest has been even partially cleared or logged. Removal of the trees not only alters the internal microclimate of the forest by exposing its heretofore protected, moist, shaded interior to the sun but also leaves behind large accumulations of woody material that are exposed to the sun's drying heat. Daily temperatures soar in the deforested areas by ten to fifteen degrees Fahrenheit, which causes the woody fuels to dry and become extremely flammable.

Now, it is not a matter of *if* the area will burn, but instead of *when* it will burn. The ultimate result is a quick, dramatic change from a dense, closed-canopy forest virtually immune to fire to a weedy, flammable pasture in which fires are common and often occur repeatedly—to the exclusion of a new forest![151]

For all of the above reasons, the professional caretakers would be wise to commission synthesis reports from the nation's forest ecologists in order to determine what today's most productive and sustainable conditions are in terms of the forest as a biological living trust in their respective bioregions. To accomplish this ecologically sound restoration of our public forests would require an imaginative use of the various silvicultural techniques, as well as an ecologically suitable role for fire.

Fire in Western Forests

Public forests will *not* remain healthy if treated with the "hands-off" policy that many environmentalists are fighting for in their current push for no logging in national forests. Having altered the fire regime in forests over the last century, we have not only set them up to burn in catastrophic conflagrations but also set them on a course of becoming something other than the healthy, fire-resistant, old-growth forests they would have been under the governance of Nature's ecological principles. Should a Douglas fir forest in western Oregon be left alone, meaning no logging, it would, without the intervention of fire, become a climax forest of western hemlock and western red cedar. The worst thing we can do is become so enamored with our view of today's forests that we eliminate the very processes that both created them and will change them.

Consequently, to save them, we must re-think our stance based on the latest scientific understanding of how disturbance regimes design, maintain, and govern ecosystems. Then, bolstered with such information, we must act wisely, boldly, and within the limitations of Nature's ecological principles to emulate the disturbance regime(s) each forest was following prior to the mandate of suppressing fires—*none of which can happen with a hands-off policy.*

Well, you might ask, if a hands-off policy is not a workable solution, what is? How does one take the vision of the Forest Council, as guided by the Council of Children, and transform it into policy and action? One does so in three primary ways: (1) *replacing* the *economic reasons* for "managing" public forests as a commodity *with ecological reasons* for treating them as an integrated, living system—a biological living trust; (2) asking whether a decision and its proposed action will serve the purpose of furthering the vision or hinder it; if the latter, the decision and proposed action are altered accordingly or discarded; and (3) passing each decision that in any way addresses policy and actions on public forests through the four gates of trusteeship:

- At the first gate, the trustee will ask: "Is this decision necessary to the ecological integrity of the forest over time?"
- At the second gate, the trustee will ask: "Is this decision economically sound in the long term?"
- At the third gate, the trustee will ask: "Is this decision directly and indirectly beneficial to the welfare of my co-workers and the public, present and future?"
- At the fourth gate, the trustee will ask: "Is this decision something I want to be remembered for?"

None of this means that the commercial aspect of forestry on public lands is to be ignored. To the contrary, it is clearly part of the duty of the professional caretakers to see that any treatment of forests on public lands has as viable an economic component as possible—while meeting the intent of the vision and the ethics of trusteeship in the present for all generations, neither of which can be accomplished with the suppression of fire.

Fire is a physical process through which Nature originally designed forests in the Western United States.[152] But that's not how Gifford Pinchot, first chief of the U.S. Forest Service, saw it as he rode through park-like stands of ponderosa pine along the Mogollon Rim of central Arizona in the year 1900.

It was a fine day in June as Pinchot rode his horse to the edge of a bluff overlooking the largest, continuous ponderosa pine forest in North America. It was warm, and everything seemed flammable. Even the pine-scented air seemed ready to burn. What a sight! Sitting on a horse in a sun-dappled, perfumed forest without a logging road to scar the ground, without a chainsaw to tear the silence, to simply behold such a forest.

"We looked down and across the forest to the plain," he wrote years later. "And as we looked there rose a line of smokes. An Apache was getting ready to hunt deer. And he was setting the woods on fire because a hunter has a better chance under cover of smoke. It was primeval but not according to the rules."[153]

The forest over which Pinchot gazed on that June day in 1900 was three to four hundred or more years old, trees that had germinated and grown throughout their lives in a regime characterized by low-intensity, surface fires sweeping repeatedly through their understory. These fires, occurring every few years or so, consumed dead branches, stems, and needles on the ground and thinned clumps of seedlings growing in openings left by vanquished trees. Although fire had been a major architect of the park-like forest of stately pines that Pinchot admired, he didn't understand fire's significance in designing the forest or the indigenous peoples' roll in perpetuating them.

What's interesting is that *Sunset Magazine* contained an article in 1910 that recommended to the fledgling Forest Service that it use the Indigenous American's method of setting "cool fires" in the spring and autumn to keep the forests open, consume accumulated fuel, and in so doing protect the forest from catastrophic fire.[154] Unfortunately, that recommendation came the same year that, in the space of two days in "Hell," fires raced across three million acres in Idaho and Montana and killed 85 firefighters in what is called the "Big Blowup." It would be ten years after the Big Blowup before fires in Western forests and grasslands were effectively controlled.[155]

For decades thereafter, the U.S. Forest Service was dedicated to putting all fires out. By 1926, the objective was to control all fires before they grew to ten acres in size. And a decade later, the policy was to stop all fires by 10 a.m. on the second day.[156] Such a policy is misplaced, however, because it ignores the primary cause of forest fires.

The response of the Forest Service is not surprising when one considers that most people prefer the devil they know to the devil they don't, which is but saying that the "terrible known" (a catastrophic fire) is often more comfortable than the unknown (setting "cool fires" like the Indigenous Americans), even if the unknown promises to be better. People thus chart a course by consciously avoiding charting a course, which means that a manageable situation is neglected until it is thoroughly out of hand.

Given enough time without human intervention, virtually all forest ecosystems evolve toward a critical state in which a minor event sooner or later leads to a major event, one that alters the ecosystem in some fundamental way. As a young forest grows old, it converts energy from the sun into living tissue that ultimately dies and accumulates as organic debris on the forest floor. There, through decomposition, the organic debris releases the energy stored in its dead tissue. In this sense, a forest equates to a dissipative system in that energy acquired from the sun is dissipated gradually through decomposition or rapidly through fire.

Of course, rates of decomposition vary. A leaf rots quickly and releases its stored energy rapidly. Wood, on the other hand, generally rots more slowly, often over centuries in moist environments. As wood accumulates, so does energy stored in its fibers. Before its suppression, fires burned frequently enough to generally control the amount of energy stored in accumulating dead wood by burning it up and so protected a forest for decades, even centuries, from a catastrophic, killing fire.

Regardless, a forest eventually builds up enough dead wood to fuel a catastrophic fire. Once available, the dead wood needs only one or two very dry, hot years with lightning storms to ignite such a fire, which kills parts of a forest and sets them back to the earliest developmental stage, the herbaceous stage. From this early stage, a new forest again evolves toward old age, again accumulating stored energy in dead wood, again organizing itself toward the next critical state, a catastrophic fire that starts the cycle over.

In this way, a 700-year-old forest that burned could be replaced by another, albeit different, 700-year-old forest on the same acreage. In this way, despite a series of catastrophic fires, a forest ecosystem could remain a forest ecosystem. And that is why the old-growth forests of western North America have been evolving from one major fire to the next, from one critical state to the next.

Although Pinchot knew about fire, he was convinced it had no place in a "managed" forest, so fire was to be vigorously extinguished, especially after the 1910 "Big Blowup." In addition, conventional wisdom dictated that ground fires kept forests "understocked," and more trees could be grown and harvested without fire. Further, surviving trees, like the ones Pinchot had seen in Arizona, were often scarred by the fires, and this kind of injury allowed decay-causing fungi to enter the stem, hence reducing the quantity and quality of harvestable wood. Finally, any wood not used for direct human benefit was thought an economic "waste."

It was in part Pinchot's utilitarian conviction about fire's economic evil that became both the mission and the metaphor of the young agency he built. Here we must keep Pinchot's two ideas in mind: fire has no place in a "managed" forest, and what is not used to the material benefit of society is an "economic waste."

In Pinchot's time and place in history, *he was correct and on the cutting-edge,* especially since the ecological problems caused by such thinking were unbeknownst to him. Nevertheless, incorporation of these ideas into forestry began to take their toll. Only now, after decades of suppressing fires and planting blocks of even-aged trees across whole landscapes, has the significance of the resulting changes in the composition, structure, and function of forests become evident.[157]

Recent evidence shows that some ponderosa pine forests in northern Arizona had only 23 trees per acre in pre-European times. This pre-European density is in stark contrast to the current density of approximately 850 trees per acre, with predominantly small diameters.

The increased density of trees is estimated to have caused a:

- 92 percent drop in the production of grasses and other herbaceous plants
- 31 percent reduction in stream flow
- 730 percent increase in accumulated fuels on the forest floor
- 1,700 percent increase in volume of sawtimber
- decrease from 115 to -8 in the index of scenic beauty
- habitat shift from open, savannah-like conditions in pre-European times to dense forest[158]

Since the advent of fire suppression, there has been a general increase in both the number of trees and the amount of woody fuels per acre. There has also been a decrease in the extent of quaking aspen, which often resprouts from roots following fire, and a corresponding increase in those species of trees that tolerate shaded conditions under closed canopies. And some of these shade-tolerant trees have grown into the forest canopy and formed a ladder of fuel up which a fire can burn from near the ground to the tops of large trees.

It may seem odd, but the ecological degradation of the ponderosa pine forests in northern Arizona in recent times is because of *too many trees*. Such increased tree density was caused by the introduction of livestock grazing and suppression of fire that shifted the open, park-like, pre-European forests of huge stately trees to dense, closed-canopy stands of less-vigorous young trees—an entirely different forest ecologically.

In addition to a rapidly increasing body of historical evidence of the long-term presence of fire as a creative dynamic in forests, we are learning that many plants have special adaptations to fire, even physiological and ecological requirements for fire. Ponderosa pine, as I said before, often has thick bark that insulates its living stem tissue from intense heat; further, as bark reaches a certain temperature, bubbles of resin within it explode, casting tiny smoldering pieces away from the trunk, an effective mechanism for reducing a dangerous buildup of heat.

While fire's role in the physiological and ecological requirements of individual species of plants may be relatively clear, there is greater difficulty in determining how fire regimes design whole forests. Most historical studies, termed "chronosequences," (= time sequences) are hampered by effects of unknown events that can result in erroneous interpretations of data. These unknown events make it particularly important to study major ecological processes in an integrated fashion because such mechanisms are interdependent. As well, the variability in fire regimes is more likely to be important to plant communities than are the mean values computed from some arbitrary period of fire history, as exemplified by the fact that unusually long periods without fire may lead to establishment of fire-susceptible species.

With respect to variability, the simultaneous occurrence of such fire-free periods and wetter climatic conditions may also be extremely important to such species as ponderosa pine that have episodic patterns of regeneration (occur as specific, discrete episodes), as opposed to plants whose regeneration patterns are continual. Consequently, while statistical summaries of fire histories are useful in a general comparison of fire regimes in different forests, the influence of fire on a particular ecosystem is strongly historical. Some forests are more a product of unusual periods of climate and fire frequencies in the short term than they are of cumulative periods of climate and fire frequencies in the long term. This kind of relationship points out that nothing in Nature is designed by the averages of anything, a concept made clear by the sleuth, Sherlock Holmes, in his comment about people's behavior.

In his discussion of the predictability of human behavior, Holmes touched the core of special cases and common denominators. He saw each person as an unpredictable special case, but if enough special cases are studied with an eye for their common traits or common denominators, then certain predictions can be made about their behavior.

Swiss psychiatrist Carl Jung put it differently. Because knowledge is a matter of getting to know individual facts, theories help very little. The more a theory claims universal validity, the less capable it is of doing justice to individual facts. Any theory based on experience is necessarily statistical and formulates an ideal average that abolishes all exceptions at either end of the scale and replaces them with an abstract mean. Although the mean is valid mathematically, it does not necessarily occur in reality. Nevertheless, the mean is an unassailable, fundamental component of the theory. Exceptions at either extreme, although equally factual, do not appear in the final result because they cancel each other out.

Take pebbles. Determining the weight of each pebble in a bed of pebbles and finding an average weight of 145 grams reveals little about the real nature of the pebbles. If, on the basis of these findings, you try to pick up a pebble weighing 145 grams on the first try, you would be in for a serious disappointment. Indeed, it might well happen that no matter how long you search, you would not find a single pebble weighing exactly 145 grams.[159]

These same concepts are used in forestry, but we who deal with forests and forestry do not derive as much meaning from them as did Sherlock Holmes and Carl Jung. Consider, for instance, that each sale unit of timber (special case) is measured to derive the volume of marketable wood fiber that can be cut for a profit. Because the volume of each tree is averaged, an abstract mean is produced and an absolute dollar value is assigned to each sale, which is valid precisely once.

Once the original old-growth stand is cut, a linear-minded forester expects to plant more trees and derive *at least the same volume* of wood fiber from a greater number of much younger trees on those same acres (the predictable, artificial, absolute minimum) in perpetuity. Whether or not this goal is achieved depends on the "cooperation" and "coordination" of all the ecological variables that constitute the forest—including climate.[160]

As long as we think of each stand or sale unit as an isolated special case, we will never understand the forest because a special case is thought to be out of relationship and out of context to the whole, which is an aggregate of special cases. We must understand how each delineated stand of trees relates to neighboring stands, to the water-catchment, and to the landscape before we can begin to understand the forest and make any kind of reasonable predictions about future trends in its behavior.

Let's look at a concrete example. In the late 1970s, the supervisor of the Wallowa-Whitman National Forest in northeastern Oregon wanted to cut more old-growth trees, but he legally had to maintain a given number of old-growth trees per acre. His challenge was that the old growth grew in discontinuous clumps and along streams in relatively remote canyons, a legacy of old cutting patterns and current topography, both of which denied him the ability to cut more of the old trees. Therefore, he had the number of old-growth trees counted and *averaged* over the acreage of the forest, which, on paper, would allow him to cut more. In his eagerness to circumvent the law/policy, however, he overlooked the fact that much of the land he had averaged the trees over was thinly soiled, rocky meadows where no trees could grow. Hence, all an average does is hide the truth concerning variation, such as the kinds of fire that create a forest.

With respect to fires in the western coniferous forests, there were two general kinds prior to the cumulative effects of fire suppression having crossed the ecological threshold in the 1950s between forest maintenance fires and forest replacement fires (*photo 11*). Forest maintenance fires were those that burn frequently enough—every few years or decades—to dissipate the accruing energy stored in dead wood on the forest floor but which did not get intense enough to kill the forest. These fires actually allowed ancient forests to develop by "fire proofing" them in a way that protected them for centuries against the vagaries of fire seasons.

Forest replacement fires, on the other hand, were infrequent—every few centuries—and killed much of a forest, thereby starting it over. These fires were dependent on unusually hot, dry summers and such things as "dry lightning storms" to ignite them, but once ignited, they became "fire storms" that razed the ancient forests.

Although it's possible that climatic change could account for the increased numbers of so called "wildfires" since the onset of fire suppression, changes in forest composition, structure, and function are the most likely cause. Intensive study of historical fires in the ponderosa pine forests of the American Southwest

11. The forest replacement fire in this photograph, like typical forest-replacement fires, left most of the dead trees standing as snags, which directly belies the often-stated notion by the timber industry that a clear-cut *in any way* emulates a forest fire. (Photograph by Dave Conklin.)

prior to 1900 has failed to document any cases wherein fires killed a forest by burning through treetops. In contrast, there have been numerous fires since 1950 that have exceeded 5,000 acres and totally razed the forests down to mineral soil. The intensity of these fires is attributed to the amount of woody fuels on the forest floor and to dense stands of young trees within the forests—both of which have come about since 1900. This said, I must caution that fires burn differently in the coastal redwood forest of northwestern California than they do in the higher-elevation, lodgepole pine forests of northwestern Colorado. And the Douglas fir-western hemlock forests of northwestern Oregon burn differently than the ponderosa pine forests of northern Arizona.

It is now 2002, and four events are converging at this point in a way that may alter the future of public forests:

First, a decision made roughly a century ago—the suppression of forest fires at any cost—has prepared forests to burn, just as they have been burning across the Western United States for the last few years. In 2002, for example, fourteen major fires were burning in Oregon alone as of July 30th, and, as of that date, had scorched about 350,000 acres.[161] By August 7th, the acreage burned was almost 507,000.[162]

The Biscuit Fire, which began on July 13th with a lightning strike in the southwestern corner of Oregon, was already, by the 10th of August, the largest fire on record in Oregon in more than a century—and still out of control, at which time, having burned over 330,000 acres, it had the undivided attention of more than 6,000 firefighters. By August 21st, the fire was fifty percent contained at just over 471,000 acres, with a crew of 6,607 people, and a cost of $84.5 million. (Jim Furnish, retired Deputy Chief of the U.S. Forest Service for the National Forest System, told me that $84.5 million is the annual budget of about eight average national forests.) As August 24th dawned, the fire had burned just over 492,000 acres, had a perimeter of 206 miles, and was 65 percent contained. August 28th came as the fire reached just over 500,000 acres, at which point it was ninety percent contained at a cost of $108.8 million. On August 29th, the Biscuit Fire, still only ninety percent contained, ceased to grow; it now stood at just over 500,000 acres (over 463 square miles in extent) and a cost of $115.6 million to fight.

On the 6th of September, two months after it started, the Biscuit Fire was declared to be contained at just over 500,000 acres—but was not to be fully controlled until put out by autumn rains. The fire had a perimeter of 405 miles of roads, bulldozer lines, and hand lines, and sported a price tag of $133.1 million. Finally, on the 8th of November, after more than two inches of rain fell, the Biscuit Fire was declared to be totally under control. The fire had a cost of $154, 868,016, to which another $6,982,633 had been spent on replacing road culverts and seeding burned hillsides.[163]

Second, the number of people with homes or cabins in the fie-prone forests of the West, as well as in other parts of the country, is substantial *and increasing*. While in Taos, New Mexico, some years back, I visited a cluster of homes tucked into the forest. The access roads were exceedingly narrow, and the dense forest was within easy reach of the houses, but the people refused to have any trees cut. They had created a death trap for themselves, a situation that brings to light the tension among the ecological necessity of fire, forest regulations for the sake of public safety, and the rights of private property.[164]

The dilemma of self-created fire traps was exemplified at a meeting in Woodland Park, Colorado, where a woman was heard to yell, when warned by an official from the U.S. Forest Service of the increasing danger of fire to her home: "Don't tell me not to build a home in the mountains! I don't want to hear that!" Whereupon, a Forest Service employee was heard to mutter: "Then don't ask us to put the fires out."[165] Here, human emotion clouds both reason and ecological reality, but there is a solution—create an "at your own risk" zoning ordinance that dictates personal liability with no recourse to fire insurance, federally subsidized or otherwise, for anyone situating their home within a public forest or immediately abutting one. (The same should be true for anyone situating their home or cabin within a floodplain in or adjacent to public lands.)

The *third* converging event is the misguided notion on the part of many preservationists that no commercial logging of any kind, which they express as "no cut," should be allowed to take place on public lands. Although misguided ecologically, having worked for 12 years for the U.S. Bureau of Land Management and with the U.S. Forest Service, I can understand where this belief came from, since both agencies have long been "in bed" with the timber industry. The "no-cut" attitude will, nevertheless, continue to prepare the public forests for increasingly catastrophic fires that, in turn, will be harder to control and consequently will both cost more in terms of federal and state tax dollars that could be better spent elsewhere and will destroy ever-more private property. This scenario may well be exacerbated by the *fourth* converging event—global warming, whatever the cause.

Because of the dynamic nature of evolving ecosystems and because each system is constantly organizing itself from one critical state to another, we can only caretake an ecosystem for its possible evolution, not for a sustained yield of products, a point eloquently made by American scientist Alfred Lotka: "It is not so much the organism or the species that evolves, but the entire system, species and environment. The two are inseparable." Ergo, the only "sustainability" for which we can caretake is whatever best ensures an ecosystem's ability to adapt to evolutionary change (such as global warming) in a way that may be favorable for us humans. But what might global warming mean with respect to forests?

While it is true many scientists agree that global warming is a reality, no one knows if it will be slow enough to allow forests, and ecosystems in general, to adapt with a minimum of disruption or so fast that they become increasingly challenged by rising temperatures, outbreaks of insects and diseases, and an increasing frequency of fire. Should the latter occur, forests along the edges of the Great Plains may well be forced to retreat even as the grasslands expand, a scenario that has happened repeatedly since the close of the Wisconsin Glaciation, some 10,000 years ago. In addition, lowland forests will move upward in elevation and north in latitude, whereas high-elevation forests may cease to exist on some mountaintops.

This notwithstanding, there is good evidence that forests wax and wane as interglacial periods come and go, including ours, the Holocene. After catastrophic disturbance, such as the beginning of an interglacial period, there is a phase of ecological development (progressive succession) that results in changes in the soil, increases in productivity, available nutrients, and biomass (which is the combined weight of all living things in an area). Following the peak of this progressive succession, and in the absence of another major disturbance, a regressive succession begins that reverses the process with a decline in productivity and biomass due to a growing impoverishment of the soil that is characteristic of the later phases of an interglacial stage.

These progressive successional stages are characterized by: rising temperatures and the development of diverse, light-demanding vegetation growing in "new," mineral soils followed by high temperatures and the development of a closed-canopy forest still growing in fertile soils. Forests not subjected to catastrophic disturbance over a period of some thousands of years enter a period of regressive successional stages characterized by: decreasing forest cover and deteriorating soils followed by declining temperatures, increasingly open-grown vegetation, and infertile soils. Combining the progressive and regressive successional stages as they span an interglacial period can be characterized thusly: rising temperatures and the development of diverse, light-demanding vegetation growing in "new," mineral soils—>high temperatures and the development of a closed-canopy forest still growing in fertile soils—>decreasing forest cover and deteriorating soils—>declining temperatures, increasingly open-grown vegetation, and infertile soils.

The decline in biomass is accompanied by an increased limitation of phosphorous in relation to nitrogen and a reduction in the release of phosphorous from decomposing plant material, reduced decomposability of the litter from the dominant species of plants, as well as reduced respiration by the soil, and a reduced ratio of bacteria to fungi. These relationships (coupled with changes in climate and biotic interactions, as well as a myriad others) all point to the relative health and fertility of the soil as the driving factor in the long-term patterns of succession within a forest.

I will now put some of this interglacial succession into motion by recounting a brief history of vegetational changes in the United States over the past several thousand years, which will give global warming a historical context. As the last glacial stage, the Wisconsin, reached its maximum development, about 70,000 years ago, there was a gradual reduction in the average temperature of the world, ranging from a five-degree Fahrenheit reduction at the equator to a ten-degree reduction at 35 to forty degrees north latitude, about the latitude of the state of Kansas, and a 25-degree reduction at the edge of the ice sheet. The snow line was lowered about 3,000 feet for each ten-degree drop in average annual temperature; consequently, the boreal forest was in South Carolina and sub-arctic plants and animals occurred as far south as Virginia, Texas, and Oklahoma.

Between 20,000 and 15,000 years ago, the Appalachian highlands were open, treeless, grass and sedge tundra with permafrost; whereas the environment in Virginia was dominated by white spruce and is thought to have been a spruce parkland interspersed with ponds, marshes, and prairies. Spruce trees were also scattered throughout much of what are now grasslands, from southern Saskatchewan southward to northeastern Kansas, and from western Minnesota southward to western Missouri. Between 16,600 and 13,700 years ago, both spruce and tamarack occurred in southwestern Missouri. In Texas, on the other hand, pine was dominant over spruce about 17,400 years ago.

The center of the North American continent was a grassland during the Pleistocene epoch. The borders of the grassland expanded or withdrew as temperatures waxed and waned, and with each change, the grassland competed across zones of contact with coniferous forest, deciduous forest, or desert shrub. During the interglacial times and warmer parts of glacial cycles, the deciduous forest retreated eastward, and the western grassland occupied the abandoned area. With cooling temperatures and subsequent increases in precipitation, the forest reoccupied its former distribution while the grassland retreated westward. Members of two principal plant communities (northern and southern) within the grassland itself migrated, mingled, competed, and evolved. Many species were lost, while new ones may have come into existence.

Changes from coniferous to deciduous forest took place rapidly following the recession of the Wisconsin Glaciation. Forest composition changed first in Georgia approximately 13,600 years ago. It took another 1,000 years for change to occur in the higher and more northern Appalachians. Since the initial coniferous-deciduous forest turnover, the resulting habitat has been a closed-canopy, oak-dominated forest with only minor changes in climate and composition of forest species.

The warming, drying post-Wisconsin climate saw the eastward retreat of the deciduous forest, as well as an eastward and presumably southward movement of the grassland. Spruce disappeared from Texas about 12,500 years ago. The eastern deciduous forests became isolated from the western coniferous forests soon after the end of the Wisconsin glacial stage, approximately 10,000 years ago, and the intervening area evolved into the Great Plains of the Midwestern United States and south-central Canada.

The picture is somewhat more complex along the western and northern portions of the North American grasslands. Here, the grasslands were more or less mixed with trees and shrubs that followed an east-to-west climatic gradient of decreasing precipitation and humidity. The broad-leaved deciduous forest of the East gave way to open, drier coniferous forests dominated by pine and juniper in the West. As the climate continued to warm and dry, wind-driven grass fires began to play a significant role in the evolution of the grasslands. Fires burned unchecked until quenched by rain or halted by abrupt differences in topography. As fires swept repeatedly through savanna-like grasslands, the trees and shrubs died out and were replaced by grasses.

Some fire-sensitive trees survived, however, not only along rivers and streams but also on protected escarpments and rocky promontories that served as natural firebreaks. Nonriparian, scarp-restricted woodlands extended across the Central Plains in the latitude of the Prairie Peninsula — from Indiana through Illinois, Iowa, and Nebraska to Wyoming. Numerous wooded scarps also occurred throughout the flat monotony of the short-grass steppe from eastern Montana southward to New Mexico and Texas.

Recurrent drought and subsequent fires prevented the scarp-restricted trees from regaining a hold in the grasslands. Although the trees originated as part of a regional forest, prairie fires helped to maintain the open conditions by consuming the seasonally dry grasses as they swept across the great expanses of flat topography. Abrupt topographic features stopped the fires and protected scattered islands of trees. So, although climate was a factor in the evolution of the grasslands in the center of the North American continent, so too were the vastness and flatness of the Great Plains and the annual, fire-carrying dieback of the grasses.

Northern plant and animal communities of eastern North America are composed largely of post-Wisconsin glacial-stage plants and animals that immigrated to terrain previously stripped of living organisms by glacial ice because competition favored species adapted to harsh northern environments that were capable of rapid dispersal. Animal communities on the southern edge of the glacier were composed of northern and temperate species. Unadaptable temperate species continued to inhabit local refugia (areas of unaltered habitat) while those with various degrees of adaptability survived in various unglaciated areas.

During the Wisconsin Glacial Stage, a large lake filled the Fort Rock Basin of south-central Oregon, east of the High Cascade Mountain Range. The habitat was a mixture of grassy plains, riparian woodlands, and water. The grassy areas of the lowland glades and upland prairies were occupied by two species of horses and three species of camels. The stream valleys, with strips or clumps of woodland, were suited for the large ground sloth, mammoth, two species of peccaries, and a bear, while the streams themselves had beaver and muskrats. These animals were hunted by dire wolves and by people, who are thought to have moved into the basin about 11,000 years ago. The lake and streams contained five species of carps and suckers as well as Chinook salmon. All these nonhuman species, except the salmon, are extinct. Presence of the salmon indicates a stage of overflow through an outlet to the Pacific Ocean that allowed the salmon to reach Fort Rock Lake. The salmon became land-locked when the overflow ceased, but persisted in the lake until the end of its existence about 10,000 years ago.

The 10,000-year interval between the end of the Wisconsin Glacial Stage and the beginning of historic times must have been of crucial importance in the establishment of modern patterns of vertebrate distribution throughout the intermountain west. Prior to 7,000 years ago, white-tailed jackrabbits lived in the lower elevations of the present Fort Rock Basin. The white-tailed jackrabbit is adapted to the colder climates of higher, more northerly regions and tends to occupy grassy habitats. Sage grouse, elk, and bison, each with similar habitat affinities, all lived in the Fort Rock Basin. In addition, the pika or rock rabbit lived in jumbles of broken rock.

About 7,000 years ago, the climate began to change. As the climate warmed and dried, between 7,000 and 5,000 years ago, the plant community shifted from one that was primarily grasses and herbs to one that was primarily shrubs. This shift in plant communities caused the local extinction of the white-tailed jackrabbit, sage grouse, pika, elk, and bison. The mountain cottontail and black-tailed jackrabbit, on the other hand, increased as the habitat changed because both are adapted to the warmer climates of lower, more southerly regions and tend to occupy shrubby habitats.[166]

What might global warming mean for humanity today? It means that, among many other unknowns, the amount and timing of precipitation is a "wild card" that foresters may someday have to contend with.

Today, a major task facing scientists, "managers," and caretakers is to re-educate themselves and the public. Smoky Bear (not to mention Walt Disney's Bambi) has done a resounding job of convincing people that forest fire is a terrible and wasteful thing. The new message must be that, while fires set by careless people can be needlessly destructive, under specific conditions fires are both beneficial and necessary to the long-term ecological health of North American coniferous forests.

Those who love forests must remember that *change* itself is the only *constant* feature in a forest. Each forest evolves in response to short- and long-term variations in its environment, such as climate, over which humans have no control. We, the materialistic species, can neither arrest nor control changes wrought by Nature through the illusionary omnipotence of our knowledge; but if we could, we would grossly and irreversibly upset the tenuous relationships through which Nature has produced the very forests we value and want to protect.

It now seems obvious that the effort to eliminate fire from our forests is an economic choice made by humans that became a rule of conduct. And one great obstacle to changing this rule is our illusion of knowledge, which economic imagination draws with certainty and bold strokes, while scientific knowledge advances slowly by uncertain increments and contradictions.

The challenge for today's forestry profession is to sit humbly in the forest, there to discover the rules by which forests have evolved and lived out the millennia. In fact, were the farsighted Gifford Pinchot alive today, knowing what we now know about fire, he would undoubtedly say: "Fire in managed forests is primeval *and* according to the rules. We must therefore learn all we can about its role in order to manage western forests for the benefit of our forests and our citizens." (By managed forest, I mean a forest that is modified and manipulated to accomplish specified objectives.)

The "patterns across the landscape," of which I wrote earlier, were created in the Western United States primarily by the waxing and waning of fire since the end of the Wisconsin Glaciation, some 10,000 years ago. That is why I think it wise to emulate the patterns created over the centuries by fire as the template for caretaking the public forests. Further, as far as I know scientifically, the coniferous forests, shrublands, and grasslands of the West are all adapted to cope with fire. Because knowledge of fire, its dynamics, and patterns are the model it would be wise to follow, a fire ecologist must be assigned to the staff of every District, where the landscape was historically designed by fire, in order to ensure the proper use of fire in the caretaking of our public forests.[167]

Fire Patterns Across the Landscape

The fire patterns that can still be traced today, both on the ground and from the air, show that fires were "opportunistic" in their burning and so left a mosaic of habitats.[168] This mosaic was created because a given fire would burn intensely in one area, coolly in another, and moderately in still another, all of which depended on what kind of fuels it encountered, how large they were, how dry they were, and how they were arranged. By "arranged" I mean whether there was dead wood lying horizontally on the ground or small, live trees that formed ladders of explosive fuel as they reached into the crowns of the large, old trees under which they were growing. To give you a clear image of how fire behaves, I have included a few paragraphs from my book "Forest Primeval." The story begins with a fire in the year 987:

It is hot and dry, very dry. The sun glides toward the Pacific Ocean. Swallows and bats mingle briefly in twilight before changing insect patrol over the top of the 700-year-old Douglas fir forests. An owl hoots. Then another. Black clouds, like great stalking cats, devour the moon, and the night grows still and heavy.

A breeze begins to stir the treetops, like gentle probing fingers. Rain begins to polka dot bare, parched soil on ridge tops with punctuating "plop, plop, plops." The wind grows strong, becomes urgent, and the rain begins to hiss as the wind scurries hither and yon through the forest canopy. Lightning flashes, thunder cracks and rolls, and the odor of ozone fills the air. Lightning slashes the darkness. Silence. Thunder. Lightning. Silence. Thunder. Lightning strikes the top of a Douglas fir and spirals down its trunk. Lightning strikes another tree and another. Then, with a simultaneous ear-splitting crack, lightning spirals down a 250-foot tall Douglas fir and strikes the ground igniting the forest floor at the base of the tree.

A small fire now casts its flickering light against the trunk of the old tree. The fire grows and spreads faster and faster across and up the slope. It finally comes to a jumble of five fallen Douglas firs that died of root rot and blew over in a storm three years ago. Here western hemlock, the shade-tolerant tree that grows under the Douglas fir and replaces it in old forests, forms a ladder of various ages and heights from the forest floor into the tops of the giant firs. As the flames begin to burn more intensely and leap higher from the fallen trees, they reach part of the hemlock ladder, which seems at first resistant to the heat, then explodes in flames. The flames climb the ladder into the tops of the firs. They roar through the treetops creating their own wind and irresistibly devouring the forest before them. The irregular line of fire 250 feet above the ground throws great sheets of flame 300 feet into the air that appear to disconnect themselves from the fiery torrent in an effort to defeat the darkness. Leaping, exploding, darting forward, the flames bridge streams and open spaces and start new fires ahead of the advancing inferno. The immense shooting flames roar into the night under the twinkling light of silent stars.

In other parts of the forest, the fire creeps along the ground burning twigs and branches, pausing occasionally to consume a snag or partly buried fallen tree. It races uphill and creeps along the contours, flares and smolders, is cool in some areas and reaches over 1,200 degrees Fahrenheit in others, kills all vegetation and spares individual trees, small groups of trees, and whole islands. And along the big river in the valley, the great fleeces of dry moss growing on the trunks of the trees act as fuses made of the old black gunpowder up which the flames shoot into the trees, igniting their crowns like torches. Thus Nature alters Her canvas that She may create a new forest.

As the pale light grows in the east on the 5th day of August 987, tall, blackened, smoking columns appear against the sky. Large fallen trees still smolder. Here and there are the charred bodies of deer that could not outrun the raging inferno, and under a large blackened tree is the body of an Indian hunter who was trapped by the flames and killed by the falling tree. A sudden, muffled "thump" punctuates the silence as a large, standing, dead tree or snag, weakened by age and the fire, crashes to earth.

The sun rises; the day becomes hot and still. Smoke hangs like a pall. By late afternoon, a light breeze off the ocean 100 air miles away begins to blow the smoke eastward up the west flank of the Western Cascade Mountains. The breeze stirs the fire in a smoldering, fallen tree and it erupts in flame, its brightness a seeming mirror reflection of the stars as darkness settles quietly over the land.

August becomes September then October, and on the 13th of October, it becomes cloudy as the storm front off the ocean reaches the mountains. Warm, moist, blustery wind heralds the storm as falling rain magnifies the odor of burnt forest. It rains all night. Cloaked in the darkness, the surviving Douglas fir trees begin to shed their seeds, for as the tree originally lay dormant in the seed so the seed lies dormant in the cone of the tree. But now the blustery wind shakes the giant firs and loosens the seeds in their ripe cones hanging from boughs over 200 feet above the ground. The winged seeds whirl and spin as they ride the wind various distances to earth.

Twenty-three years later (1010): The burn, now 23 years old, is a vast mosaic of habitats arranged over the landscape according to Nature's combined patterns of soil types, slope, aspect, elevation, moisture gradients, and severity and duration of the fire, including areas the fire missed. Many of the south-facing slopes are shrub-fields dominated by snowbush; other areas, where fire scorched the soil, are only now in grasses and herbs. Streamsides are growing up with alder, willow, and vine maple, among other species. Where Douglas fir seeds fell and germinated along the edge of the forested areas in 988 and 989, single trees, small clumps, and dense thickets are scattered around the burn.

The plant community is changing in some areas of the shrubfields. Snowbush, with nitrogen-fixing bacteria associated with its roots, has been adding nitrogen to the soil. In addition, its roots harbor some of the same mycorrhizal fungi that are symbiotic with Douglas fir and thus have kept the inoculum alive in the soil. As soil conditions change, and the offspring of snowbush cease to survive, spaces become available between parent shrubs, and where spaces and Douglas fir seed come together, seedlings are becoming visible. And so the burn progresses from bare ground, to grasses and herbs, to shrubs, to young forest.

Although the fire of 987 killed many trees, it destroyed very little wood. Many snags of fire-killed trees stand like black and silver spires and columns of various sizes and shapes, as regal sentinels against time, a measure of the burn's renewal. A few snags are already surrounded by thickets of Douglas fir, as though they are trying to hide but can't quite cover all of themselves.

Many of the snags already have cavities that were excavated over the years by various woodpeckers before and after the fire. Some cavities, high up in tall snags, are used for nesting by tree swallows; lower cavities are used by western bluebirds. And amongst the solid snags are a few hollow ones that are open to the top. Because of their scarcity, these hollow snags are a premium habitat in the burn as they are in the ancient forest.

Fifty years later (1037): On the 19th of June, lightning starts a fire in the ancient forest at the northwestern edge of the burn, near the big river. The fire creeps around the forest floor consuming twigs, branches, and now and then large, fallen trees that are dry enough to burn. One large, almost buried, fallen tree near the edge of the ancient forest smolders below ground for days before it finally reaches mineral soil and burns out. This low-severity fire also removes fire-sensitive species, such as seedlings and small saplings of western hemlock and western red cedar, as well as some of the smaller Pacific yew trees. The fire not only adds to the overall diversity of the forest but also "fire proofs" this portion of it to some extent by removing the easily combustible fuels on the ground.

The giant Douglas firs are not injured by the fire because the outer bark at their bases is 10 to 20 inches thick, which protects their living tissue from the heat, and their crowns are far too high above the flames to sustain damage.

When the fire reaches the burn, however, it climbs quickly into the trees at the edge of the young forest because their living limbs form a ladder from the ground up. Once in their crowns, the fire races up the mountain, burning unchecked until heavy rains from a thunderstorm on the 22nd of June, followed by an overcast day, and more rain on the 24th puts it out. The fire has blackened a strip that is a mile wide when it reaches the scattered clumps of trees in the subalpine forest.

The fire affects nutrient cycling throughout the forest. Nutrients, such as nitrogen, are converted to gas as the vegetation burns and are lost into the atmosphere; some are lost in the ashes that become wind borne and leave the site on the huge blasts of heat from the fire; still others combine with oxygen and so become different compounds.

Once the fire is out, more nutrients are lost from or redistributed within the burned area by wind that blows the ashes around and, later, by rain and melting snow that leaches some of the nutrients out of the soil and carries them into the streams and into the big river and ultimately into the sea.

Not all of the nutrients are lost, however; rain replaces a little of the nitrogen and in a small way balances the account. In addition, some of the nutrients that remain will be more readily available to the plants soon to inhabit the burn.

The new burn's loss of nutrients becomes some other area's gain because the nutrients are all eventually redistributed within the water-catchment, the landscape, the geographical area, the continent, and the world. Nothing is truly lost, only removed for a time. And someday, some other area's loss will be the new burn's gain.

Stream flows increase strikingly after the fire has killed most of the vegetation. Runoff from the melting snow begins earlier in the spring of 1038, and the runoff peaks are higher. The fire not only has blackened the surface of the soil, causing it to increase dramatically in temperature on sunny days, but also has altered its behavior. Blackened soil absorbs heat and water therefore evaporates more readily. Where the soil is severely burned, it becomes more repellent to water than before the fire; so rather then infiltrating deeply into the soil, the water runs off at or near the soil's surface.

Water temperature increases immediately after the fire because stream channels, now devoid of protective vegetation, are exposed to direct sunlight. But not all vegetation along the streams is killed. Some of the shrubs live below the surface of the soil and will resprout by early summer of 1038 and, within a year or two, will again shade the water of the first-order streams.[169]

Remember the description of the 500,000-acre Biscuit fire? Well, it's a graphic, concrete example of how fire, like the one described in my story, alters Nature's canvas to its own design. Satellite images revealed that nineteen percent (9,600 acres) of the area within the fire's perimeter had not burned, and 41 percent (205,000 acres) had burned at low intensity—leaving live, green trees, while clearing out areas that were overgrown with vegetation in the understory due to decades of fire suppression.

Of the remaining forty percent of the area (191,500 acres), about 22.6 percent (113,000 acres) burned with moderate intensity, clearing the ground of dense understory vegetation, and is thought to have killed most—but not all—of the trees *without* burning up their needles. Yet if one were to look at the base of the intact needles, one would likely find them to still be alive at the core. Although one of the largest fires in modern Oregon history, it severely burned *only* 15.7 percent (78,500 acres) of its total acreage, wherein it left behind little more than ashes and charcoal. In other words, the Biscuit fire behaved just like the scientists who study fire expect forest fires to behave; it burned in a mosaic pattern that created a diversity of habitats in time and space across the landscape.[170]

In effect, the Biscuit fired did exactly what needed to be done—it thinned, and thus helped to "fire proof" a choked forest, which is analogous to an observation the Greek Paracelsus made long ago: namely, that "the deity which brings the illness also brings the cure." That is, a short-term fire is often a long-term cure for itself because, having consumed the energy stored in the dead wood, it has greatly reduced the probability of the same acres burning again in the near future. In fact, charred rings on the old trees indicate that the dry areas of the Biscuit fire historically burned about every 50 years, whereas the wetter sites burned about every 70 years.

I spent some time in October, 2002, looking at fires that burned that previous summer, six years earlier, 36 years earlier, and 120 years earlier. Each fire, no matter how intense, left almost all trees standing, even those that were only three inches in diameter. In addition to the wonderful diversity of habitats that a single fire creates within a given area, successive fires over variable years compound the vast mosaics of interconnected macrohabitats as they design and redesign landscape-scale patterns over decades and centuries, creating patterns that are readily discernible on aerial photographs. In so doing, fires maintain and revitalize ecological diversity.

"Thinking and acting at the landscape level tends to ameliorate many forest health concerns," says Jim Webb, emeritus forest supervisor with the U.S. Forest Service. Jim goes on to say that "Our innate human inability to maintain cohesive, long-term programs, combined with our short attention spans and *issues-*

du-jour mentality, predisposes us to worry about issues that are really not that big a deal in the grand scheme of things." This does *not*, however, mean that forest fires can simply be ignored.

The Misguided Fear of Fire

Although the misguided fear of forest fires has plagued a number of presidential administrations, the Bush administration is the most current. Be that as it may, the George W. Bush administration, in a knee-jerk reaction, has asked Congress to speed up the efforts to prevent forest fires by exempting ten million acres of public forest from normal, environmental reviews *and* by eliminating all administrative appeals on projects to thin forests, where the risk of fire is high. This maneuver would force any challenges by the public to a given thinning project directly into court, where judges would be prohibited from issuing temporary restraining orders or preliminary injunctions to stop such projects. Such a constraint means—if the timber industry is true to its old pattern—that the commercially valuable, old-growth trees can continue to be cut while the court decides if the project is even legal.[171]

Here, *the real environmental challenge* is not the prevention of fire, but rather *saving the large, old-growth trees*, which are *not* the ones targeted to be cut in a thinning—the young, small, suppressed ones are. Should Bush be granted what he wants, the timber industry can cut as many old-growth trees as the Forest Service will allow—without an environmental review, without any public involvement in the decision governing the caretaking of the federal forests held in trust for that same public. In other words, the people in the timber industry could cut with impunity virtually any tree they chose.

To my way of thinking, the Bush administration's attitude is both arrogant and a travesty. And it reminds me of a comment made by Ralph Waldo Emerson: "Your attitude thunders so loudly that I can't hear what you say." Besides, it took a century of mismanagement by *both* Democrats *and* Republicans to get our public forests in the sad shape they are in today; they cannot be "fixed" overnight. It will take time and patience—*something we Americans need to practice*.

That notwithstanding, there are some people, such as Dr. Dennis P. Lavender, a retired professor of old school of exploitive forestry, who believes that herbicides are *better* than fire in *managing* a forest and, consequently, should replace fire, even though western forests have been "designed" by fire over the centuries and millennia. To make his point, Lavender cites all the traditionally perceived, short-term negative effects of fire, to wit:

> 1. Much of the brush has evolved to resist fire. It will sprout vigorously after burning, and dormant seeds—which require exposure to heat to germinate—will result in increased stands of brush, creating a Sisyphean task [= endless task].
> 2. Fire will destroy the many invertebrates required to build soil, together with their food. True forest biodiversity is below the soil surface, as

there are at least 1,000 species of such organisms per square yard.

3. A map of Oregon forest soils shows that low nitrogen levels coincide largely with areas that have been repeatedly burned. Even supposedly harmless ground fires can be very destructive to vital soil processes and the soil (not trees or other plants or vertebrates), which is our basic resource. The use of fire that can vaporize the majority of soil nitrogen is not compatible with sustainable forestry.

A procedure preferable to fire would involve any one of a range of chippers and the use of herbicides. The above is ecosystem specific.[172]

Let's examine this linear-minded approach to forestry one point at a time:

1. *"Much of the brush has evolved to resist fire. It will sprout vigorously after burning, and dormant seeds—which require exposure to heat to germinate—will result in increased stands of brush, creating a Sisyphean task."*

What Lavender says is true in *some* areas (such as northwestern California, where the tenacious tanoak can occupy a site for decades *if* the site is *repeatedly burned*), but in other areas (such as many parts of the Intermountain West) fire is needed to keep groves of quaking aspen from dying out due to a *lack of sprouting*, which is a current and serious threat to the local biological diversity, especially since aspen groves are natural clones. The outcome of both of these examples, as well as all other cases, depends on a variety of circumstances, such as the intensity and duration with which a fire burns on any given site; besides, the "brush" Lavender objects to is also part of forest succession and landscape-scale diversity. What Lavender is really concerned about is "management-created competition" of "brush" (= shrubs) with "crop trees," something that does not equate to sustainable forestry, unless "forestry" is defined as having nothing but crop trees on all acres all of the time.

But then, forests *do not* operate in the same linear trajectory that linear-minded foresters perceive them to. Nature's forests go through "autogenic succession." If you remember, autogenic succession can be characterized by "successional stages," a concept that refers to the developmental stages that a forest goes through from bare ground to an old-growth forest, but not in nice, discrete steps.

2. *"Fire will destroy the many invertebrates required to build soil, together with their food. True forest biodiversity is below the soil surface, as there are at least 1,000 species of such organisms per square yard."*

Lavender is again correct, but forest fires, even ones, such as the Biscuit fire, rarely burn so intensely overall that they leave nothing but scorched, "dead" soil. Instead, forest fires—as exemplified by the Biscuit fire—create mosaics of habitat diversity on a landscape scale due to the variable intensities with which they burn. I bring this up because landscape-scale diversity is critical to both the health of the forest as a whole and to the maintenance of "genetic stepping-stones" that may one day be critical to the genetic ability of a forest to migrate up in elevation or north in latitude in the face of changes in the global climate.

Consider that large fallen trees partly buried in the soil tend to retain their inner moisture and so can, and often do, act as refugia that allow the tiny organisms necessary to the governance of the soil to survive the flames and heat and reoccupy the surrounding soil when conditions become favorable again. As a side note, one of the things I was never able to do was study the ecological relationships of casehardened wood or charred wood in terms of forest processes. As far as I know, this research still needs to be done.

3. "*A map of Oregon forest soils shows that low nitrogen levels coincide largely with areas that have been repeatedly burned. Even supposedly harmless ground fires can be very destructive to vital soil processes and the soil (not trees or other plants or vertebrates), which is our basic resource. The use of fire that can vaporize the majority of soil nitrogen is not compatible with sustainable forestry.*"

If by "sustainable forestry" Lavender means a continual cover of "crop trees" on all acres all of the time, then, by his definition, he is correct. To view a "forest" in this way is to attempt to make the crop trees into an independent, economic variable—a physical impossibility.[173]

To me, there is much more to a forest as a living system than simply one crop of trees after another. As I think of sustainable forestry, it can be constructed on two premises: (1) within some limits, a forest will persist, provided the existing disturbance regime is compatible with its continued existence and (2) given the chance, a specific condition within an ecosystem, plant community, or successional stage will recur in an approximation of its predecessor. This means that by accepting the first premise, the second is allowed to fulfill itself, but not on all acres all of the time and perhaps not even on our timetable all of the time.

Bear in mind that we can gently guide Nature, but trying to force Nature will surely result in resistance, unwanted outcomes, and situations over which we have no control. Nature, after all, is not our servant, but rather, given a chance, may become our partner.

The first part of Lavender's last statement in his number 3 above (*The use of fire that can vaporize the majority of soil nitrogen is not compatible with sustainable forestry.*") is true. Fire does "vaporize" the nitrogen in the material that is burned, but that does not mean nitrogen is absent from the soil. Besides, successional plants, such as lupine, snowbush, and red alder, readily replenish soil nitrogen where it has been depleted.[174] Omitted from the above statement is that exploitive forest management also tends to deplete nitrogen from the soil the crop trees are growing in.[175]

Finally, "*A procedure preferable to fire would involve any one of a range of chippers and the use of herbicides. The above is ecosystem specific.*"

First, chipping large woody debris homogenizes it in small pieces that destroy its function as habitat for those organisms, such as nitrogen-fixing bacteria, that perform critical processes within the dead wood itself. In addition, creatures that are vital to forest health because they disperse viable spores of ectomycorrhizal fungi, such as red-backed voles and northern flying squirrels, depend on the large, dead wood for habitat—habitat that chipping destroys. Chipping also destroys the water-holding capacity of large woody debris. In essence, chipping as a means of controlling the amount of flammable fuels on the forest floor in the name of "fire management" renders it all but impotent as functional biological capital to be reinvested in the health of the soil. In this case, I am using the term "biological capital" to include the diversity of organisms that use large, decomposing wood on the forest floor as habitat and food; the diversity of processes these organisms perform; and the large wood as the "storage organ" of elements, water, organisms, and processes that allow nutrients to be cycled from the old trees through the soil into new trees, all part of a biologically sustainable forest.

Second, despite our current level of knowledge, no one is or can be certain of what happens to chemicals once they are introduced into the environment, be it through intensive farming or intensive forestry, because once we have introduced something into the environment, it's almost immediately out of our control. Such introductions include the estimated thirteen million pounds of pesticides applied each year in Oregon alone, which it seems no one in authority really wants to monitor. Nor does anyone know where the chemicals travel, how they behave in their journey, or how long their effects persist.[176] To help you understand what I mean, let's consider how chemicals affect salmon, which are highly *adapted*, unlike humans who are highly *adaptable*. Here, as with the bull trout discussed earlier, adaptable means the *ability* to adapt to change, *adapted* is a *fait accompli* with respect to given circumstances, so they are not the same thing. Being highly *adapted* means salmon are rigid in their habitat requirements.

On top of all the obstacles (dams, improperly placed culverts, degraded habitat, etc.) that salmon must today face in their journey from the forest to the sea and back again, modern humanity has added the effects of herbicides, fungicides, insecticides, rodenticides, fertilizers, gasoline, oil, and other pollutants to the once clean waterways the salmon of old knew as their domain. Since the salmon's life cycle is a chain with many links, the diverse continuum of habitats required to fruitfully fulfill their life functions, from the tiny forested streams wherein they spawn and hatch all the way to the sea, makes saving the salmon a process of locating all the broken links in the chain of their life cycle and fixing them again.

For example, a primary problem with pollution that affects salmon comes from intensive agriculture—*including intensive forestry*—that laces the soil with both toxic pesticides and synthetic fertilizers, thereby killing fish and other marine life, while damaging marshlands in more than a third of the nation's coastal areas, according to the National Academy of Sciences. This problem means that every autumn, Monte Graham gets nervous because, as a soil and water conservation officer for Marion County, in western Oregon, he knows that tons of farmland soil—and its myriad synthetic chemicals—will erode and leach into ditches and streams with each inch of rain, where the chemicals can cause bone deformities in the baby salmon, damage their reproductive systems, destroy their food supply, and block their adaptation to saltwater. The chemicals can also prevent migrating adults from finding their home waters to spawn. Not surprisingly, a study by the U.S. Geological Survey found that the Willamette River Basin of western Oregon to be among the most degraded in the nation, due in part to chemical runoff from cities, farms, *and* intensively managed forests.

The persistence of many pesticides in the environment has "sub-lethal" effects on biota and is a possible culprit in the escalating crisis of salmon in the Pacific Northwest. *Some pesticides break down into, or combine with, other compounds that are even more toxic than the original chemical combination.* Under certain conditions, tiny amounts can accrue high toxicity. In addition, the only part of a pesticide that is tested for toxicity is the active ingredients, which usually form a tiny portion of the solution. This means the larger, untested portion of most pesticides, containing other, so-called "inert" chemicals, can be even more toxic than the active ingredients.

Chemicals are a potential problem that we cannot afford to overlook because their effects on fish are seldom either visible or directly lethal, according to Jim Martin, whose career as a fishery biologist spanned thirty years, many with the Oregon Department of Fish and Wildlife. Martin's concern is that focusing on the chemical effects as they relate to people may eclipse their possible effects on salmon. After all, says Martin, "salmon are a lot more sensitive than people to water quality. They have to meet the most stringent survival requirements imaginable."[177] While salmon are adapted to the effects of forest fires, they are *not* adapted to the environmental effects of chemicals, including herbicides used in exploitive forestry.

Emulating Fire Patterns

With the preceding as background for the thinking embraced in exploitive forestry, I believe it would be wise to have those folks who deal with fire examine satellite images and aerial photographs of all the fires that burned in 2002 and 2003, as well as those that have burned over the last decade. From these data, they could characterize the fires in terms of patterns of burn at the landscape-scale, water-catchment scale, and stand scale.[178] The purpose of this exercise would be to determine the common denominators and differences among the fire patterns and begin to figure out how various portions of a fire pattern, or even an entire fire pattern, could be emulated in caretaking the federal forests *without having* to use fire per se on all acres all of the time. This said, however, both prescribed fire and letting some fires burn as Nature intended would be part of the caretaking strategy because fire is the most authentic emulation of itself.

Whereas it seems obvious that Nature's fires are going to be less and less tolerated in future times due to objections from the growing human population,[179] we must find a way to emulate the disturbance patterns whereby fire has created and maintained forests through time for the expressed reason that forest have become adapted to cope with the stresses of fire.[180]

While the literature leaves little doubt that fuels can be modified in ways that affect the behavior of fire, the best modification of fuels is *with fire*. A number of empirical studies demonstrate the effectiveness of prescribed fire in altering the behavior of "wildfires." In terms of caretaking our public forests, therefore, it seems prudent to locate areas in which fuels are modified in accord with well-constructed, experimentally driven designs that will provide the kind of knowledge necessary to emulate forest fires at an acceptable level of intensity and rate of spread for the sake of forest health and the maintenance of Nature's required ecological services.[181]

Once the common denominators and differences among the fire patterns have been determined, a conference could be convened for all the scientists in the U.S. who in any way study fire. They could spend the first days presenting what they know about forest fires and their effects on land and habitats. Following the presentations, they could break into small groups and spend a day or two discussing their impressions of the data presented and committing their conclusions to paper for a comparative discussion. The conference could conclude with a two- to three-day exploration of the questions that need to be asked: scientific, social, and economic.

The participants could then be sent home with the request to design long-term (100-year or more) experiments in caretaking public forests by emulating fire and its patterns that could be carried out in five or six bioregions in various parts of the Western United States—with a strong component of long-term monitoring, an activity that typically is the weakest link in long-term research. These designs would use a variety of means of treating a forest, such as human effort, machine, and controlled fire, and they would require the cooperation and coordination of several administrative units in each bioregion. This combined interaction would necessitate a shared vision as the participants determine how to caretake the bioregion's public forest(s) as a biological living trust.

As results are gleaned from the on-going experiments, those that are promising can be tested throughout the types of forest for which the experiments were designed. Caretaking the public forests as a biological living trust must be adaptive and adaptable to a continual, free flow of new information and the permission to experiment with new ideas—including permission to fail, a vital and necessary ingredient of success.[182]

Just as the suppression of fire as a practice of exploitive forestry is adversely affecting the entire American West, so are exploitive forestry practices adversely affecting the nation's supply of water.

Water-Catchments

Seventy-five percent of the surface of the Earth is covered with water, but more than ninety-seven percent of it is salt water that makes up the oceans. Another two percent is frozen in glaciers and the polar ice caps, leaving only one percent of all Earthly water available in usable form for life outside of the oceans.

Water is a non-substitutable requirement of life. Its source and capacity for storage are finite in any given landscape. Fresh, usable water, once thought by non-indigenous peoples in the United States to be inexhaustible in supply, is now becoming scarce in many parts of the world. In the western United States, for example, water pumped from deep underground aquifers is today such a valuable commodity that it is often referred to as "sandstone champagne."

The availability of water throughout the year will ultimately determine both

the quality of life in a community and thereby the value of real estate. Consequently, our nation's supply of quality water is precious beyond compare. In fact,
water is the *most valuable* commodity from our public forests. But then, is water
really a commodity in the sense of economic markets or is access to water part of
the global commons—the birthright of every individual?

Is water to become the ultimate economic/environmental club with which
we bludgeon one another? This question is appropriate here because we are running out of available supplies of potable water.[183] The only solution is an environmental one—sound ecological caretaking of water-catchments on a landscape scale
that first and foremost nurtures the health of soil and water, lest everything else
become unhealthy. Like migratory birds and anadromous fish, environmental
crises know no political boundaries. Soil, water, air, and climate form a seamless
whole, the thin envelope we call the biosphere, and the biosphere is all we have
in entire the Universe.

How We Think About Water-Catchments

As with any problem, there are solutions, but we tend to think about and
look for solutions only where the symptoms are obvious, a situation seldom apparent with water-catchments. The problem with water-catchments normally begins with the headwaters, the first-order stream and its water-catchment, usually
a trickle far removed from human habitation. A first-order water-catchment is
always a special case; in fact, it is probably the only part of the land in which the
hydrology has any semblance of ecological integrity since it is the headwaters
and so controls the initial water quality for the whole catchment basin.[184]

Our thinking, and so our view of the world, is generally limited to a kaleidoscope of special cases because we choose to focus on "discrete" parcels of land as
"real estate." If we deal only with special cases, such as a mile of stream, we
perpetuate our inability to understand that particular mile of stream, the entire
stream, and the water-catchment as an interdependent whole. If, on the other
hand, we deal with a particular mile of stream (a special case) in relation to the
whole water-catchment (the common denominator), we enhance our ability to
understand both the mile of stream and the water-catchment as each is defined
by its relation to the other. Understanding how a reach of stream relates to the
whole water-catchment is like understanding how a single chair relates to a room.
(A "reach of stream" is the visible portion of a stream between two bends in the
stream.)

If you were to stand in the doorway and survey a room, you would see the
chair both in the room and in relation to the room, but when you focus only on
the chair you can no longer see the room or the chair's relationship to it. Unfortunately, most people do not see that the first-order water-catchments (headwaters)
are the initial controllers of water quality for supplies of domestic water. For this
reason, exploitive forestry allows logging to occur down to the edges of both
first- and second-order streams, even in municipal water-catchments, because
the timber is thought to have greater immediate economic value than the water.

Moreover, since politically important fish, such as salmon and steelhead, do not live at the high elevations in which most of these small streams occur, the water is deemed to be of no visible, economic importance. The invisible importance of the water in a water-catchment, far from the tap that dispenses it, becomes visible only when the water reaches human communities and becomes usable. But first, it must be stored somewhere to be available when needed.

The Hydrological Continuum

Water is a physical necessity of life. Water is perhaps the most important commodity when it comes to the sustainability of a community, so a community's supply of quality water is precious beyond compare.

The amount and quality of water available for human use are largely the result of climate and strategies for taking care of the ecological health of water-catchments. In North America, sustaining the health of water-catchments is particularly important in order to protect the annual snowpack from which the vast majority of all usable water comes. Yet protecting the quality and quantity of society's water supply is not a primary consideration of the linear-minded folks in timber corporations, so a primary goal of caretaking public forests must be to protect and, when possible, to enhance a forest's capacity to store water in the form of snowpack.

People seldom realize that drinkable water comes predominantly from forested water-catchments. Even much of the prehistoric ground water that is pumped to the surface for use in agriculture came from forested water-catchments. Salmon, water, and hydroelectric power are forest products just as surely as is wood fiber.[185]

As a nation with once bountiful resources, the United States has rarely faced limits to those natural resources.[186] Although times have changed, continuing trends and experience indicate that "informed denial" is rampant in that every additional drop of water conserved by one segment of the public is thought to be available for ever-more economic growth by another segment of the public, further raising the demand for more water and more economic growth—like a circular firing squad.[187] Effective caretaking of water will necessitate attention to both demand *and* supply.

The availability of water also depends on such variations in components of the hydrologic cycle as precipitation, evaporation, transpiration, infiltration, and runoff. Because these components are interrelated, a change produced by technology in one component of the cycle will inevitably affect other components.

In the short history of the United States, there have always been more lands and more resources to exploit and a philosophy that technology could supplement natural resources, or even supplant them as needed, an idea confidently stated by L. C. Everard, editor of the 1920 Agricultural Yearbook:

> As a Nation we have always stood on our own feet and felt ourselves masters of our own destiny. Our immense and varied natural resources

have enabled us to maintain this position and have justified this feeling. It is largely because of our confidence in the sufficiency and permanency of these resources that we have been in the past and are now able to look the future calmly in the eye and go on our way steadily improving the quality of our national life. We have always been able to look beyond the frontier of cultivation to new and untouched fields ready to supply the landless farmer with a homestead and to meet the growing demands of the country for food, clothing, and shelter. The untouched reserve has about disappeared. We have another reserve, however, as vast as that which lay before the pioneers in the old days. It is the grain and meat, the wool and the wood, the thousand and one other products of field and forest that we can add to our store by applying more intensively on the farm and in the forest the scientific principles and methods to come forth from laboratory, sample plot, and experimental farm. As the days go by we learn more and more the underlying causes of success in agriculture, we perfect methods of applying the new discoveries, we reduce more and more the element of chance and guesswork, we grow in knowledge of how to get more and better crops from the land and how to market them where they will do the most good. The answer to the problem of both producer and consumer lies in the extension of our efforts in these directions, in the use and distribution of what we have on the basis of more complete knowledge, and in putting the idle land to work and making all the land work to better purpose.[188]

Now, in 2005, we have a different picture. Today's perceived dilemma is one of stretching such resources as water to accommodate the continuing economic growth of the western United States, while protecting the existing patterns of water use, a behavioral norm requiring levels of technical development that are increasingly damaging ecologically and no longer feasible economically. Moreover, few people realize that only a small part of the water used in the United States goes to towns and cities. The overwhelming share is wastefully used for irrigation.

In California, where growth in the human population has been virtually unlimited, such growth was possible because, for many years, the "excess" water from the Colorado River was given to the state. That came to an end in 2003, when California had to give up enough water from the Colorado River to supply roughly 1.4 million people in order to ensure allocations of water for six other Western states. This reduction amounted to thirteen percent of the total water that California had been taking from the river.

The state could have avoided the cutback, if water agencies in Southern California had resurrected a deal aimed at reducing the state's over-dependence on the Colorado River. The deal called for a transfer of some of the Imperial Valley's massive share of the river's water from the valley's desert farms to the city of San Diego by the end of 2002. Even so, the Imperial Valley water board, the overseer of the nation's largest irrigation project, refused to sell a drop of water.

Compounding the problem is the fact that the Colorado River burst through a farm dike in 1905 and flowed unimpeded for more than a year into California's southeastern desert, where it filled an ancient seabed and created the Salton Sea. Runoff from farms in the Imperial Valley, that annually sees the use of a trillion gallons of the river's water, have kept the sea alive ever since. Now, with supplies of fresh water running low in the American West, the Salton Sea (largely ignored until 2002) has become the center of controversy among the competing interests in the fragile ecosystem that has supported millions of birds and other wildlife for 100 years, a region in which farming produces much of the nation's winter-time vegetables, as well as fast-growing cities.

The idea of precious water from the Colorado River collecting in an agricultural sump 227 feet below sea level is not well received by those people in Western states for whom continual growth is their focus. These people, from such states as Arizona and Nevada, are ill content to keep the Salton Sea alive if it means foregoing the potential of more water for human consumption and hence continual economic growth. To such people, saving the Salton Sea is simply a "waste" of water—a loss of potential economic gain.

For residents of the Imperial Valley there is an additional concern. Should the Salton Sea begin to dry up, it might unleash a dust storm much like the one residents of the Owens Valley complained about after the city of Los Angeles made its infamous water grab in the valley ninety years ago.[189]

According to Professor Luna B. Leopold, the persistence of the pro-economic expansion bias of the U.S. Bureau of Reclamation is increasingly inexcusable. This attitude is still held in spite of the obvious strain on both the quality and quantity of the water supply.

Although caretakers of public forests are, for the most part, familiar with the hydrologic cycle, which continues for better or for worse, the idea of a hydrologic continuum is not so familiar. A hydrologic continuum, as used by Leopold, implies the maintenance of a quasi-equilibrium, operational balance among the processes within the hydrologic cycle that involve air, water, soil, biosphere, and people. In other words, if withdrawals of water by humans are balanced with Nature's capacity to replenish the water used, the use of water can be measured in such a way that the available, long-term supply is protected from being over-taxed.[190]

There are four possible options in caretaking the use of water: the first being to discipline ourselves to use only what is necessary in the most prudent manner. Another is to protect the health of an entire water-catchment and so the supply of snow. A third is to simultaneously account for the first two, and the fourth is take water for granted and use all we want with no discipline whatsoever (as we do now through continual economic expansion) and then wonder what to do when faced with a self-inflicted shortage, as is happening in northern China.

Tianjin, a city in northern China, has sunk more than six feet in the past two years, which has damaged buildings and pipelines, as well as concentrating salt and other chemicals in the groundwater. The subsidence is caused by a growing number of funnel-shaped areas—more than thirty—beneath the North China Plain, an effect of increased pumping of the groundwater for agricultural and household uses. It is feared that all the funnels will eventually coalesce to undermine an area of 15,400 square miles.[191]

By using all the water we want in a totally undisciplined manner, we are insensitive to both the care we take of the water-catchments and the speed with which we mine the supply of stored available water.[192] As stated by Professor D. J. Chasan, "One might suppose that people would automatically conserve the only naturally occurring water in a virtual desert, but one would be wrong. Land and farm machinery have capital value. Water in the ground, like salmon in the sea, does not. Just as salmon are worth money only if you catch them, water is worth money only if you pump it."[193] Consequently, we are pumping ground water, and we are damming, diverting, and channelizing the rivers to "tame" and "harness" their water for short-term use based on poor economics, rather than nurturing the water-catchments to ensure the availability of an adequate long-term supply of water.

In fact, a curious thing happens when water flows outside the forest boundary: we forget where it came from. We fight over who has the "right" to the last drop and pay little attention to the supply—the health of the forested water-catchments.

With the growing realization of the ecological interdependency among all living forms and their physical environment, it can hardly be doubted that even "renewable" resources (such as water[194]) show signs of suffering from the effects of society's unrelenting, materialistic demands for more. These demands have degraded the renewability of resources in both quality and quantity. Water can be thus characterized because it is increasingly degraded by soil erosion, increases in temperature, pollution with chemical wastes, salts from irrigation, and overloads of organic materials.[195] Is it any wonder that the hydrological system is under stress?

The Storage of Water

The storage of a community's water originates in the soil of high mountains far from the tap you turn to fill a glass with this most precious of liquids. Water is stored in four ways: (1) in the form of snowpack aboveground; (2) in the form of water penetrating deep into the soil, where it flows slowly belowground; (3) in belowground aquifers and lakes; and (4) in aboveground lakes and reservoirs.

Most water used by communities comes first in the form of snow, either at high elevations or northern latitudes, where it melts and subsequently feeds the streams and rivers that eventually reach distant communities and cities—rivers, such as the Columbia, Snake, Colorado, Missouri, and Mississippi. Snowpack is aboveground storage that, under good conditions, can last as snowbanks late into the summer or even early autumn.

How much water the annual snowpack has and how long the snowpack lasts depends on six things: (1) the timing, duration, and persistence of the snowfall in any given year; (2) how much snow accumulates during a given winter; (3)

the moisture content of the snow—wet snow holds more moisture than dry snow; (4) where the snowfall accumulates in relation to shade and cool temperatures in spring and summer, e.g., under the cover of trees and on north-facing slopes *vs.* open clear-cuts and south-facing slopes with no protective shade; (5) when the snow begins to melt and the speed at which it melts—the later in the year it begins melting and the slower it melts, the longer into the summer its moisture is stored; and (6) the health of the overall water-catchment.

Although the first five points seem self-evident, the last one requires some explanation. In dealing with the health of water-catchments, one must consider those of both high and low elevations. How we treat our high-elevation forests (and those at more northerly latitudes) is how we treat a major portion of the most important sources of our supply of potentially available water, which originates as snow.

Snow disappears in two ways: sublimation and melting. Sublimation means that snow, accumulating in such places as the upper surfaces of coniferous boughs above the ground, evaporates and re-crystallizes without melting into water. When snow sublimates, it bypasses any role in our supply of available water. Melting snow, of course, is a different story.

As a youth, in the days before the High Cascade Mountains of Oregon were dissected by roads and scarred by clear-cut logging, I enjoyed the Spring Equinox in the high ancient forest, where I camped in snow that was fifteen or more feet deep, not counting the drifts that easily exceeded twenty feet in depth. With the advent of late spring and early summer, the snow began to melt, and the water gradually infiltrated the soil until every minute nook and cranny was filled to capacity with the precious liquid that was all the while obeying the unrelenting dictates of gravity as it journeyed along the ancient geological path toward the streams and rivers of the land on its way to the sea from whence it came. As gravity pulled the water downward through the soil, the slowly melting snow continually filled the void left by the departing liquid. In this way, the melting snows of winter fed the streams of late summer and autumn, thereby bringing water to all the forest-dependent industries human communities depend on, including the communities themselves.

Then came the roads, chainsaws, yarders, and log trucks and the ancient forests that once protected the snowpack from the heat of the sun began to disappear. As logging roads progressively fragmented the once contiguous forest and clear-cut after clear-cut merged into gigantic, naked mountain slopes (*photo 12, page 206*), the snow melted earlier and faster than in the days of my youth and in so melting saturated the soil in a much shorter short time. Now, the water-holding capacity of the soil is often reached in late May and early June, greatly exceeding gravity's ability to pull the water through the soil into the valley bottoms and thus allow the soil to absorb all the water. The inability of the soil to absorb the great pulse of water from the melting snow in May and June causes most of it to flow over the surface of the ground, where it rushes down streams and rivers, speedily fillingthe reservoirs to overflowing and so is lost to the human commu-

12. A clear-cut in which the timber company had no respect for the first-order stream, which ultimately constitutes the quality control for the water used by people. (Photograph by Chris Maser.)

nities when they need it most, late in the year.

To help you visualize what I am talking about, consider a large log, with both ends cleanly cut off, lying across the contours of a steep slope (up and down the slope) under the canopy of an ancient forest. If the snow is deep enough, the melting water infiltrates the log at its upper and is gradually pulled downward through the interior of the log by gravity until it drips out the bottom of the cut face at the log's lower end. I have often watched this phenomenon in awe of the inexorable pull of gravity.

There is, however, a caveat to this phenomenon. If the snow is deep enough to cover the upper end of the log, it can absorb the same amount of water that drips out the bottom just as long as the supply lasts. As soon as the snow is gone, the available supply of water is cut off and that remaining in the log will eventually drip out the lower end without being replenished. Therefore, the longer the snow lasts at the upper end of the log, the longer the log can act as a conduit for the water infiltrating its upper end, passing though its length, to drip out its lower end. Conversely, the faster the snow disappears from the log's upper end, the faster the supply of water from melting snow is cut off, the quicker the log progressively dries out, even as water continues to drip out the lower end. That, too, will shortly cease because, without the water stored in the snowpack above ground to cover the log's upper end, there is no replenishment for the limited supply of water pulled through the log by gravity.

So it is, when considering the supply of water for communities, that humility, wisdom, and long-term economics dictate that some forested areas on public lands, particularly at high elevations, should not be severely cut even once for the perceived, immediate, short-term dollar value of the wood fiber. To protect such areas for the storage of water in the form of snowpack will require a drastic shift in thinking because, at present, the only economic value seen in high-elevation forests is the immediate extraction of wood fiber. Nevertheless, the roading and clear-cutting of high-elevation forests, which catch and store water, affects *all* human communities, from the smallest rural village to the largest city.

In contrast, most low-elevation water catchments, which may or may not be forested, must be much larger in area than a high-elevation catchment to collect and store the same amount of water. Although snow may not be as important for the storage of water in low elevations, the ability of water to infiltrate deep into the soil is equally important. The storage of water at low-elevation, nonforested areas is often in wetlands, subterranean aquifers, and lakes, as well as in aboveground lakes and reservoirs. Regardless of where the water-catchment is located, forest roads and urban sprawl have a tremendous effect on both the quality and quantity of water that ultimately reaches a community.

Roads, Urban Sprawl, and Water

Roads affect the quality, quantity, and distribution of water in the soil of the catchment, regardless of whether they are graveled and constructed to extract timber or paved and constructed as access to homes in a housing development. The construction and use of a road severely disturbs the soil that, in turn, increases the rate of runoff, may reduce the flow of subsurface water, and alters the equilibrium of shallow groundwater. Unfortunately, the information needed to understand the effects of a road on the regime of surface and subsurface water is still limited.

Unless water infiltrates deep into the soil of a water-catchment, it runs downhill and reaches the cut bank of a logging road or even a major highway that brings it to the surface, collects it into a ditch, and puts it through a culvert to begin infiltrating again. The water then meets another road cut, and so on. Water is sometimes brought to the surface three, four, or more times before reaching a stream. Water is purified by its journey through the deeper soil, but not by flowing over the surface of the ground. Roads bleed water from the soil the same way cuts in the bark bleed latex from a rubber tree or sap from a sugar maple.

In fact, ditches and gullies, such as those that form on the downhill side of culverts passing under roads, function effectively as pathways down which water flows. The denser the network of roads, the greater the drainage of water over the soil's surface, and the less time it takes for peak flows to occur.

This poses a question: How deep into the soil is deep enough for water to avoid the ditches at the bases of banks alongside roads? I have seen roadbeds blasted out of solid rock to depths of fifty or sixty feet, and I have seen water seeping out the "bottom " of this same rock into the roadside ditches in July and

August, a predicament symptomatic of the disruption in the flow of water. This means we are bringing precious water to the surface of the ground, where it not only evaporates but also becomes polluted by sediment, oil, and chemicals from the road's surface and human garbage in the ditch. Consequently, roads have a negative, cumulative effect on the hydrological cycle of a water-catchment and on the purity of the water that ultimately reaches human communities.

For example, a logger drains waste oil from a loader onto a landing. Where does the oil go? It soaks into the soil and is carried downhill by water and gravity. True, the oil will be diluted by the time it reaches the ditch of the road twenty feet below the landing, but it is still polluting the soil as it goes. With the winter rains of the fifth year, the oil collects in the ditch, is mixed with other oil that leaked from passing vehicles, is flushed through the culvert under the road, and continues its journey. Given enough time, it will reach the stream 100 feet below the road and pollute the stream that flows into the small reservoir that supplies the local community with drinking water.

Moreover, disrupting the flow of water through the soil on steep slopes, even forested slopes, can cause instability and increase erosion during a severe rainstorm or as snow melts. Such conditions in the vicinity of the seeping water can cause soils to become saturated with little or no infiltration that in turn weakens them and leads to greater local runoff of water over the surface of the soil and hence greater erosion.[196]

In housing developments within a water-catchment surrounding a community, on the other hand, roads and streets are paved, creating an impervious coating over the surface of the land. This impervious layer prevents the water, both rain and melting snow, from infiltrating into the soil, where it can be stored, purified, and can recharge existing aquifers and wells. Instead, the water remains on the surface of the roads and streets, where it mixes with pollutants that collect on the pavement.

Because paved roads and streets are lined with curbs and gutters, the now-polluted water is channeled from the paved surface into a storm drain. In addition, each house has an impervious roof that collects water and channels it into gutters along the edge of the roof. Upon collecting water, the gutters channel it, more often than not, out to the street, where it joins water from the street going down the storm drain. It is then conducted either directly into a sewage treatment plant or directly into a ditch, stream, or river.

In any event, the water is not usable by the local people. Beyond that, the storm water either adds to the cost of running the treatment plant, where it must be detoxified, or it pollutes all the waterways through which it flows, from its point of origin into the ocean.

The effect of roads, streets, parking lots, and the area covered by houses, all of which eliminate the infiltration of water, is cumulative. Enough roads, streets, parking lots, and roofs over time can alter the soil-water cycle as it affects a given community. Remember that the quality and quantity of water is an ecological variable, irrespective of the fact that many linear-thinking economists and "land developers" consider it an economic constant.

Even if water was a constant, a variable is introduced with construction of a single logging road. The variable is compounded by constructing and maintaining multiple logging roads to extract timber. In addition, intensive forestry, such as clear-cutting, alters the water regime that affects how the forest grows. In this way, a self-reinforcing feedback loop of ecological degradation in a water-catchment is created, altering the soil-water regime, that in turn alters the sustainability of the forest, that in turn affects the soil-water regime in a never-ending cascade of cause and effect. Eventually, the negative effects are felt in those communities that are dependent on a given water-catchment or drainage basin for their supplies of potable water.

We humans can continue to degrade the forested water-catchments and impoverish our supply of water, or we can risk abandoning our conventional thought pattern and, with a strong, concerted commitment, reverse the trend—one of the main purposes of treating public forests as biological living trusts. In the final analysis, we must remember that only so much water is available, and with a change in the global climate, that amount may become even more variable and unpredictable than it already is. That notwithstanding, more water cannot be found in a courtroom, no matter how hard we try or who holds the priority rights to the water already available. And so it behooves us all as American citizens to consider how we care for the sustainability of the stream-order continuum that constitutes the water-catchments on public lands—lest the reservoirs and wells go dry.

The Stream-Order Continuum

The stream-order continuum operates on a simple premise: Streams are Nature's arterial system of the land. As such, they form a continuum or spectrum of physical environments, with associated aquatic and terrestrial plant and animal communities, as a longitudinally connected part of the ecosystem in which downstream processes are linked to upstream processes.

The idea of the stream-order continuum begins with the smallest stream and ends at the ocean. The concept centers around the resources of available food for the animals inhabiting the continuum, ranging from invertebrates to fish, birds, and mammals—including people.

As organic material floats downhill from its source to the sea, it becomes smaller in size, while the volume of water carrying it becomes larger. Thus, small streams feed larger streams and larger streams feed rivers with partially processed organic matter, such as wood, the amount of which becomes progressively smaller the farther down the continuum of the river system it goes.

This is how the system works: A first-order stream is the smallest undivided waterway or headwaters. Where two first-order streams join, they enlarge as a second-order stream. Where two second-order streams come together, they enlarge as a third-order stream, and so on.

The concept of stream order is based on the size of the stream—the cumulative volume of water, not just on what stream of what order joins with another stream of a given order. To illustrate, a first-order stream can join either with another first-order stream to form a second-order stream or it can enter directly into a second-, third-, fourth-, fifth-, or even larger-order stream. The same is true of a second-order stream, a third-order stream, etc.

In addition, the stream-order influences the role played by streamside vegetation in controlling water temperature, stabilizing banks, and producing food. Streamside vegetation is also the primary source of large organic debris, such as tree stems at least eight inches in diameter with their rootwads attached, or tree branches greater than eight inches in diameter. Forests adjacent to streams supply wood (stems, rootwads, and large branches from trees. Erosion also contributes organic material to the stream.

Wood in streams increases the diversity of habitats by forming dams and their attendant pools and by protecting backwater areas that are important winter habitat for fish. In addition to the wood itself, habitat diversity in the streams and rivers of the western United States has been historically maintained by regular flooding, droughts, and every imaginable condition in between these extremes. The variability of the conditions experienced by the streams and rivers continually shift the wood around and alter its function in a way that augments ecological diversity in space and time, thereby causing indigenous organisms to evolve in ways that allow them to cope with the extremes of survival. Three examples are: cottonwood trees, a caddisfly, and a giant waterbug.

Cottonwood trees, which once grew in profusion along the banks of western streams and rivers, where they provided shade, woody debris, and nutrients to the aquatic-terrestrial interface, have all but disappeared to the detriment of the ecosystems they served. Cottonwoods require the bare, scoured banks that result from floods in order for their seeds to germinate and grow, despite the fact that some mortality of the trees themselves is experienced as a consequence of the flooding. Today, because of flood-controlling dams, cottonwood trees are dying out in many areas.

There is a caddisfly that inhabits a stream system in the mountains of Arizona, where it is subjected to the extremely violent force of flash floods that occasionally scour out the stream channels. The caddisfly, in turn, has evolved through the generations to metamorphose from the immature, aquatic state into their winged, adult phase during a period that is almost perfectly timed to miss the most common season of flooding, which keeps enough of the population out of harm's way to perpetuate the species.

Finally, a giant waterbug, which lives in some desert streams, has adapted over the last 150 million years to "read" the weather and make a mass exodus from a stream that is about to experience a flash flood. During the exodus, the waterbugs literally climb the canyon walls to escape the dangerous waters, but return to the stream within a day.

When rivers are "harnessed" and "tamed" with dams, the organisms that have evolved to cope with Nature's disturbance regimes are likely to die out and be replaced by a range of different organisms. The shift in habitat and the attendant aquatic organisms that result from the construction of dams can dramatically alter how the ecosystem functions in a way that is detrimental to the food web within the entire drainage basin affected by the dams, such as preventing driftwood from completing its journey from the forest to the sea.

Conversely, when streams and rivers are unrestrained, the driftwood they carry provides nutrients, a variety of foundations for biological activity, and both dissipates the energy of the water and traps its sediments.

Processing the organic debris entering the aquatic system includes digestion by bacteria, fungi, and insects that are aquatic in their immature stages (such as midges, stoneflies, mayflies, and craneflies, as well as physical abrasion against such things as the stream bottom and its boulders. In all cases, debris is continually broken into smaller pieces that make the particles increasingly susceptible to microbial consumption.

The amount of different kinds of organic matter processed in a reach of stream (the stretch of water visible between two bends in a channel, be it a stream or river) depends on the quality and the quantity of nutrients in the material and on the stream's capacity to hold fine particles long enough for their processing to be completed. The debris may be fully utilized by the biotic community within a reach of stream or it may be exported downstream.

Debris moves fastest through the system during high water and is not thoroughly processed at any one spot. The same is true in streams that do not have a sufficient number of instream obstacles to slow the water and act as areas of deposition, sieving the incompletely processed organic material out of the current so its organic breakdown can be completed. So small streams feed larger streams and larger streams feed rivers.

As a stream gets larger, its source of food energy is derived more from aquatic algae and less from organic material of terrestrial origin. The greatest influence of terrestrial vegetation is in first-order streams, but the most diversity of incoming organic matter and the greatest diversity of habitats are found in third- to fifth-order streams and large rivers with floodplains.

Small, first-order, headwater streams largely determine the type and quality of the downstream habitat. First- and second-order streams are influenced by the configuration of surrounding landforms and by the live and dead vegetation along their channels. This "riparian vegetation" interacts in many ways with the stream.

The canopy of vegetation, when undisturbed, shades the streamside. The physical energy of the flowing water is dissipated by wood in stream channels, slowing erosion and fostering the deposition of inorganic and organic debris (*photo 13, page 212*). These small streams arise in tiny drainages with a limited capacity to store water, so their flow may be scanty or intermittent during late summer and autumn, but during periods of high flows in winter and spring, they can move prodigious amounts of sediment and organic material.[197]

What I have just described is the beneficial aspect of the stream-order continuum, but there is a sinister side to this story as well, a tragically human side.

13. An old-growth Engelmann spruce has fallen across a small stream in the Cascade Mountains of western Oregon, where it slows the stream's velocity during winter's high flow, allowing gavel to collect on its downstream side (facing you); collects driftwood, which is visibly jammed up against the old tree on its upstream side; and forms winter habitat for trout, young salmon, and steelhead on its downstream side, where the fish can get out of the direct flow of the current as winter's high flow rushes over the old tree. (Photograph by Chris Maser.)

Ditches along forest roads (and elsewhere) form a continuum or spectrum of physical environments (the same as streams), a longitudinally connected part of the ecosystem in which "downstream" processes are linked to and influenced by "upstream" processes. The ditch continuum begins with the smallest ditch and ends at the ocean. So it is that little ditches feed bigger ditches, and bigger ditches eventually feed streams and rivers that ultimately feed the ocean. Remember further that as organic material (food energy) floats downhill from its source to the sea, it gets smaller—more dilute—as the volume of water carrying it gets larger.

Here the question is: What happens to the continuum concept when a ditch is polluted? To pollute a ditch means to contaminate it by dumping human garbage into it or by consciously or unconsciously discharging noxious substances into it, such as oil or hydraulic fluid from vehicles and logging equipment, both of which in one way or another disrupt biological processes, often by corrupting the integrity of their chemical interactions.

While Nature's organic matter (food energy) from the forest is continually *diluted* the further down the stream continuum it goes, pollution (especially chemical pollution) is continually *concentrated* the further down the ditch continuum it goes because it gathers its potency from the discharge of every contaminated ditch that adds its waters to the passing flow. Hence, with every ditch we pollute while caretaking and/or using our public forests, the purity of the stream and river accepting its fouled discharge is to that extent compromised, and the amount of pollution that ends up being dumping into the estuaries and oceans of the world through the stream/ditch continuum is staggering.

I say this for two reasons. *First,* I have seen ditches in North America, Europe, Asia, and Africa discharging their foul contents directly into streams, rivers, estuaries, and oceans. *Second,* in 1969, I found a population of montane voles (meadow mice to most people) living along a ditch that drained an agricultural field. The voles, whose fur was an abnormally deep yellow when I caught them, lost the yellow with their first molt in the laboratory when fed normal lab chow, whereas those along the ditch retained their yellow pelage.[198]

Even with evidence in hand, I could find no one in the Department of Agricultural Chemistry at the local university to acknowledge this color deviant, let alone examine it in an effort to find the cause—undoubtedly some agricultural chemical compound, which, if a fertilizer or herbicide, could just as easily have been a chemical compound used in exploitive forestry. Nevertheless, they all turned their backs, even when I presented them with the evidence, the live, yellow voles.

So I learned that chemical pollution in ditches is not visible to the eye of human consciousness in the flowing of their waters, but it may become visible in the sickening of the environment. And in 1984, as part of a committee called to Washington, D.C., to help the United States Congress frame the ecological components of the 1985 Farm Bill, I learned in far greater depth about the incredible non-point source chemical pollution of our nations surface waters (ditches, streams, rivers, and lakes) and ground waters (lakes and aquifers) from today's chemical-intensive agriculture, *including exploitive forestry.*

How, I wonder, can we learn to care for rivers and oceans if we continually defile the ditches that feed them? The answer is that we cannot!

We must learn to care first and foremost for the humble things in our environment, such as a roadside ditch, *before* we can learn how to care for the mighty things in our environment, such as a river. Defile the ditch, and we defile the stream, river, estuary, and ocean; protect the ditch and we protect the stream, river, estuary, and ocean in like measure. I cannot end the discussion of streams and ditches without accounting for transportation the system because, as it is currently designed in forestry, it has a dramatic effect on the quality of water.

The Transportation System

There is nothing—*nothing*—environmentally friendly about a road, any road.[199] Roads not only act as conduits for the invasion of exotic plants throughout most of North America, where they are dispersed by such things as automobiles, but also augment the illegal killing of wildlife, such as wolves in Southeast Alaska.[200] Roads serve one purpose, to allow human access to an area in a motorized vehicle of some kind, and there were already more than 385,000 miles of roads in our national forests alone by 1998! Deny vehicular access to land, especially public lands in the United States, and the outcry is almost instantaneous.

A few succinct data points from early 1998 will put the issue of roads, just on national forest lands, into perspective:

- An estimated 373,000-mile spider web of official roads crisscrossed our national forests, in addition to some 60,000 miles of "ghost roads" created by the repeated use of four-wheel drive and off-road vehicles that were not officially out there, and did not count as part of the extensive network of county, state, and national roads.
- More than 7,000 bridges existed on national forest roads.
- Commercial and recreational use of national forest roads had increased ten fold since 1950.
- Recreational use of national forests had increased to 350 million days annually and was projected to continue increasing, which increases the number of forest fires caused by human access.
- An estimated 1.7 million recreational vehicles and 15,000 logging trucks and timber harvesting vehicles traveled on national forest roads every day.
- Of the 191 million acres of national forest lands, an estimated 34 million were in wilderness status; an estimated 33 million were in unroaded blocks of 5,000 acres or more, where roads could be constructed; and eight million of the latter were classified as "suitable for timber production."
- Only about forty percent of the national forest roads in 1998 were maintained to standards that met environmental protection or acceptable levels of human safety. The cost of upgrading the delinquent sixty percent of the national forest road system was estimated to be over $10 billion. The high cost is due to an aging road system and a lack of regular maintenance, where more than three-quarters of the roads in need of maintenance are over fifty years old.
- Scientific evidence is continually reinforcing the fact that the ecological impacts of roads are more extensive than previously thought. Roads create migration routes for invasive, exotic plants and insects. They also fragment habitat and wildlife migration corridors and are a major cause of landslides and high levels of sediment in streams and rivers that is associated with the reduction in fish habitat, including that of salmon.[201]

In addressing the transportation system, I am going to share a few ideas that I think are necessary in order to caretake our public forests, as much as possible, as a seamless whole in space and time.

When a system of transportation becomes the centerpiece of a forestry operation (as it is today in most public forests), that operation is placing the primacy of its focus on the fragmentation of the forest, a focus that determines how the populations of indigenous plants and animals, as well as their habitats, are scattered across the landscape. And fragmentation of habitat has brought, and is bringing, more species to the brink of extinction than almost any other human activity.

Here, two fundamental questions might be posed: Does building more and more roads really serve the intent of sustainable forestry? Do we retain the web-like network of forest roads just because we can't remember any other way to live? These are good questions to address in caretaking public forests because there are two options in planning a transportation system: greater emphasis on ecological constraints (ecological effectiveness has primacy over economic efficiency, which tends to devalue the former) *or* greater emphasis on economic considerations (economic efficiency has primacy over ecological effectiveness).

If forest caretakers were to design a transportation system around even minimal ecological constraints, they might still be able to have a relatively good system of connected habitats, but the forest would suffer far greater fragmentation and pollution from machines, such as dripped oil, exhaust, and noise, than if *habitat connectivity as a system in and of itself* had driven the design and implementation of the transportation system. On the other hand, if forest caretakers were to design a transportation system around economic efficiency (today's policy), the connectivity of habitats as a viable system would be essentially foregone because fragmentation of habitat is inevitably maximized, as is pollution.

There is also a greater likelihood that exotic species will take over remaining parts of the landscape in a forest where the transportation system is designed around economic efficiency—as opposed to a forest wherein habitat connectivity has primacy. Fragmentation of habitat created by the primacy of a transportation system puts ever-more outside pressure on the survival of indigenous species.

In the final analysis, however, there is no one single factor as a cause for anything. All things operate synergistically as cumulative effects that exhibit a lag period before fully manifesting themselves. With this in mind, there are things that can be done to ameliorate the environmental effects of forest roads. Here are a few:

- Plan from above with aerial photographs and GIS to place the transportation system in the most environmentally effective location with respect to connectivity and quality of habitat in a contiguous forest.
- Maintain the minimum necessary system of permanent roads and so the greatest possible connectivity of habitats.
- Move existing roads out of, and away from, riparian areas and obliterate the old roadbed.
- Out-slope roads and use "rolling dips" to maintain soil stability, eliminate ditches, and thus minimize the necessity for culverts.
- Institute logging with horses, in addition to skyline and cable yarding, to minimize the use of heavy equipment to the greatest extent possible and thereby employ more people while minimizing soil compaction and pollution from machines, including noise.
- Close roads when not in use; to those of the public who complain about the lack of road access, one might point out that nothing "locks up" a forest more than roads with gates. Ergo, a minimal system of roads leaves

most of the forest open to the public; albeit, those of the public who want to visit the forest will get some exercise in the process—and they have access to millions of acres. (This said, there would have to be special accommodations made for people with disabilities.)

If you are wondering why I am now going to write about driftwood in a book on forestry, it's because of something I learned many years ago while working in Egypt. Namely, whatever we do in the environment, such as caretaking public forests, affects the whole world, and forestry does because wood travels the world's oceans.

Driftwood

I grew up in western Oregon and over the years spent considerable time at the Pacific Ocean. As a boy, I remember the huge piles of driftwood along the beaches, piles that grew or diminished with each winter storm. In fact, one of the challenges of even getting to the sandy shores of the Pacific was having to climb over the jumbled mountains of driftwood. There was so much wood, ranging from small branches to boards to whole trees, that I could build shelters from the wind that easily held fifteen or more people. Enormous piles of driftwood were simply taken for granted as part of the beach.

Then I was grown and, suddenly, the driftwood mountains were gone. What happened to them? When did they disappear and why? How could the mountains of driftwood I so clearly remember have vanished without my noticing it?

Driftwood is floating trees and parts of trees carried by water from the forest to the sea. It is a critically important source of habitat and food for the marine ecosystem, including the deep-sea floor. Even during its seaward journey, driftwood is both habitat and a source of food for a multitude of plants and animals, both aquatic and terrestrial. In addition, some driftwood becomes temporarily stabilized in a way that controls stream velocities, stabilizes stream banks, makes waterfalls and pools, and creates and protects fish spawning areas. Other driftwood protects vegetation as it encroaches on floodplains and allows forests to expand. In short, driftwood makes a vital contribution to the health of streams, rivers, estuaries, and oceans worldwide.

The natural processes by which wood disappears from streams and rivers have positive effects on the ecosystem. In contrast, the intensity and persistence of human activities, such as logging to the edge of a stream, salvage logging in riparian zones, cleaning wood out of streams (including "salvage"), and cutting firewood, have had negative effects over the last several decades. Moreover, the consequences of these actions are both little understood and far reaching.

Historically, streams replenished annual supplies of driftwood to the lower portions of river basins and out into the sea, where it washed up on beaches. And the banks of lower rivers and estuaries (the riparian corridor) were probably the common source of large driftwood in the bays.

Substantial amounts of driftwood must have been transported to the sea at the time when most riparian zones were dominated by such large coniferous trees as Douglas fir, western red cedar, and Sitka spruce and such deciduous trees as black cottonwood, bigleaved maple, Oregon ash, and red alder. Once at sea, driftwood belongs to the global commons as it travels the ocean currents. In the north Pacific, drifting trees that escape the inshore oscillations of the tidal currents enter the open ocean, where they may eventually contact the westward transport of the north Pacific gyre, a great circular vortex.

Once in the north Pacific gyre, large drifting trees can remain afloat for long periods and great distances and eventually come ashore in such exotic places as the Hawaiian Islands, where in fact trees from the Pacific Northwest account for most of the large driftwood on the beaches. Other drifted trees on the beaches of the Hawaiian Islands are indigenous to the Philippines, Japan, or Malaysia.

Anthropological records show that the beached Douglas fir and coast redwood were even integrated into the customs and rituals of oceanic cultures. Ancient Hawaiian civilizations prized the huge Douglas firs and western red cedars that washed up on their shores because local chiefs preferred them for construction of their large, double canoes, a symbol of wealth and power.

Because most driftwood discharged from tropical rivers occurs during monsoon floods, its distribution once in the ocean varies in time and space. Despite its distributional variability, ocean currents keep most of it in the coastal zone close to its source, even after a full year at sea, during which time much of it becomes waterlogged and sinks.[202]

Unfortunately, coastal deforestation in recent decades has drastically reducing the amount of driftwood available to some oceans, particularly from mangrove forests, the deforestation of which ranges from fifty percent in the Philippines to twenty percent in Malaysia.[203] Although mangrove forests may not be a major source of large driftwood, given that the generally high density of their wood causes the trees to sink, they indicate that deforestation close to coastal areas is proceeding at a pace greater than inland.

Iceland, for example, had its ancient forests already "improvidently exhausted" by the mid 1800s, according to geologist Sir Charles Lyell, who also noted the abundance and importance of driftwood:

> . . . although the Icelander can obtain no timber from the land, he is supplied with it abundantly by the ocean. An immense quantity of thick trunks of pines, firs, and other trees, are thrown upon the northern coast of the island, especially upon North Cape and Cape Langaness, and are then carried by waves along these two promontories to other parts of the coast, so as to afford sufficiency of wood for fuel and for constructing boats. The timber is also carried to the shores of Labrador [the mainland part of Newfoundland, Canada] and Greenland; and Crantz assures us that the masses of floating wood thrown by the waves upon the island of John de Mayen often equal the whole of that island in extent. [John de Mayen is now called "Jan Mayen" and is a fairly large island of Norwegian ownership lying north northeast of Iceland and east of Greenland.]

In a similar manner the bays of Spitzbergen [a Norwegian archipelago in the Arctic Ocean between Greenland and Franz Josef Land] are filled with drift-wood, which accumulates also upon those parts of the coast of Siberia that are exposed to the east, consisting of larch trees, pines, Siberian cedars, firs, and Pernambuco and Campeachy [*sic.*] woods [which means trees from Pernambuco, a state in northeastern Brazil, and Campeche, a state in southeastern Mexico on the western part of the Yucatán peninsula]. These trunks appear to have been swept away by the great rivers of Asia and America. Some of them are brought from the Gulf of Mexico, by the Bahama stream, while others are hurried forward by the current which, to the north of Siberia, constantly sets in from east to west. Some of these trees have been deprived of their bark by friction, but are in such a state of preservation as to form excellent building timber. Parts of the branches and almost all the roots remain fixed to the pines which have been drifted into the North Sea, into latitudes too cold for the growth of such timber. . . .[204]

As I said, once at sea, driftwood belongs to the global commons as it travels the ocean currents. Discharged from rivers and estuaries into the open ocean, the fate of driftwood is determined by winds and currents, which have allowed aboriginal peoples of the far north to gather tropical driftwood from the shores of the sea for fuel and building materials for summer shelters far above the Arctic tree line.

And along the coast of the Pacific Northwest, until the middle of last century, hundreds of millions of board feet of logs and driftwood entered Puget Sound and Georgia Strait from the rivers draining the Cascade Mountains of Washington and the coastal mountains of British Columbia. They were joined by large numbers of "escapees" from log rafts. Over ten billion board feet of logs were annually stored or traveled in the estuaries and the lower segments of rivers in the Pacific Northwest. A one percent escape rate would allow over 100 million board feet of driftwood to enter the ocean from this source alone.

A conservative estimate is that in decades past, as much as two billion board feet of wood per year was transported to the sea. Two billion board feet per year is a small amount when prorated across the entire North Pacific. Large driftwood, an important ecological component of Pacific Northwest streams and rivers, interfered with human activities, however, and was summarily removed. In fact, people throughout North America have systematically cleaned driftwood from streams and rivers for over 150 years.

Streams and small rivers were cleaned of driftwood from the late 1800s to around 1915, so that logs could be floated from the forests to the mills. Several "splash dams" were built on many streams to temporarily augment the flow of water in order to float logs to mills. The net effect of channel clearance and splash damming was to remove large quantities of driftwood from medium to large streams, a significant change from the conditions that formerly existed.

Over the last 100 years, millions of drifted trees and other driftwood have been cleared out of streams and rivers to facilitate navigation and reduce flooding. To this end, streams and rivers have been channelized and dammed, and marshes have been drained. In addition, most stream banks have been so altered through logging that they now have dramatically smaller and younger trees of different species than in times past.

Most big western redcedars and Douglas firs have been logged along the Cascade Mountain streams and along coastal streams greater than third order. By the 1990s, more than seventy percent of the coniferous trees greater than fourteen inches in diameter at breast height were logged to within 100 feet of fish-bearing streams on private lands.

Before the great ecological value of driftwood was known, west coast fishery managers believed that driftwood in streams restricted fish passage, supplied material for driftwood jams, and caused channels to scour during floods. Indeed, during times of flooding such fears might have seemed to be well founded, but it is now clear that the results of cleaning wood out of streams and rivers has been ecologically disastrous.

It is now apparent that neither we nor the generations of the future can afford the effects of the loss of driftwood that connects the forest to the sea and the sea to the forest. The loss of driftwood means the destabilization of streams, rivers, estuaries, complexes of sand dunes, beaches, and sand spits, as well as food webs in the oceans of the world. Sooner or later it also means the loss of commercial fishing as a way of life, because the loss of large driftwood heralds the potential decline of such fish as tuna and salmon that not only benefit from but also depend on driftwood during various stages of their life cycles.

Nevertheless, driftwood is being prevented from even beginning its journey to the ocean by the removal of as much wood as possible from the forests as a product for human consumption, lest it remain as an "economic waste." In addition, damming of rivers prevents what little driftwood even begins its journey from completing it, thereby severing the pivotal connection of the forest and the sea.

Even today, county sheriffs, port commissions, and recreational boaters still routinely clear driftwood from rivers for safety and personal convenience. As a result, most Pacific Northwest streams and rivers bear little resemblance to their ancestral conditions, when they flowed freely through pristine forests carrying their gift of driftwood to the sea.

Consequently, the supply of driftwood that is food for wood-dependent species on the bottom of the sea off the coast of North America is both dwindling and becoming more erratic. For the first time in the evolutionary history of deep-sea animals, the availability of food has become unpredictable.

If the coastal mangrove forests of the tropics continue to be destroyed through deforestation, the last direct link of the forest to the sea will be severed. The deep-sea wood-dependent species of the world will then shrink in both numbers and areas they inhabit, and some will become extinct. A question we must ask ourselves is what does extinction of some species mean in terms of the health of an ocean?[205]

In light of recent ecological information, it's imperative that we re-examine the way we treat streams and rivers and the supply of driftwood they no longer carry on its journey to the sea. Our present course impoverishes our waterways and ultimately the oceans of the world. Because healthy streams, rivers, estuaries, and oceans are critical to social benefits (and oceans are part of the global commons), it is necessary to ensure that a renewable supply of driftwood (including whole, old-growth trees) is incorporated into land-use planning and the practice of sustainable forestry on public lands as a conscious part of the biological living trust.

Chapter 7
Monitoring the Trust

Experience is the name everyone gives to their mistakes.
— English author Oscar Wilde

If you are wondering why I have waited until the end of this book to deal with monitoring, the answer is simple. "Ecology" and "economy" have the same Greek root *oikos*, meaning "house." Ecology is the knowledge or understanding of the house, and economy is the management of the house—and it's *the same house*. Clearly, one cannot fully monitor caretaking public forests by leaving out half of the equation of "caretaking."

Although the word "monitor" is variously construed, its meaning here is to scrutinize or check systematically with a view to collecting specific kinds of data that indicate whether one is moving in the direction one wishes. Monitor has the same origin as "admonition," which means a warning or caution and is derived from the Latin *monitio*, a "reminder."

With respect to caretaking public forests, monitoring means to keep watch over and warn in case of danger, such as straying from a desired course. Monitoring is to remind us of activities that we already know are too harsh and could offend the system; on the other hand, it also helps us conserve the options embodied within the system for ourselves and future generations, but this requires the ability to ask relevant questions because monitoring, after all, is dependent on questions.

The Questions We Ask

Learning how to frame a good and effective question is paramount, both for the crafting of a collective vision for the future and for the process of monitoring what action is necessary to achieve the vision. A question is a powerful tool when used wisely since questions open the door of possibility. For example, it was not possible to go to the moon until someone asked: "Is it possible to go to the moon?" At that moment, going to the moon became possible, albeit no one knew how. To be effective, each question must: (1) have a specific purpose, (2) contain a single idea, (3) be clear in meaning, (4) stimulate thought, (5) require a definite answer to bring closure to the human relationship induced by the question, and (6) explicitly relate to previous information.

In a discussion about going to the moon, one might usefully ask: "Do you know what the moon is?" The specific purpose of this question is to find out if a person knows what the moon is. Knowledge of the moon is the single idea contained in the question. The meaning of the question is clear: Do you, or do you not, know what the moon is? The question stimulates thought about what the moon is and may spark an idea of how one relates to it; if not, that can be addressed in a second question. The question, as asked, requires a definite answer, and the question relates to previous information.

A question that focuses on "right" versus "wrong" is a hopeless exercise because it calls for human, moral judgment, and that is not a valid question to ask of either an ecosystem or science. A good question would be to ask if a proposed action is good or bad in terms of caretaking a particular forest. To find out, you must inquire whether a good short-term economic decision is also a good long-term ecological decision and so a good long-term economic decision. Such questions are important because a good short-term economic decision can simultaneously be a bad long-term ecological decision and so a bad long-term economic decision, one that generations of the future would have to pay for. The point is that one must ask before an answer can be forthcoming.

In essence, questions lead to the array of options that you can choose from. Conversely, without a question, you are blind to the options. Learning about the options is the purpose of monitoring. In turn, to know what to monitor and how to go about it, you must know what questions to ask because an answer is only meaningful if it is in response to the right question.

Seven Steps of Monitoring

The crux of making a shared vision work lies in monitoring. That said, monitoring is the weakest point in the process of caretaking public forests. A shared vision for caretaking public forests must be considered a working hypothesis because it is applied to ongoing and proposed human activities within a dynamic ecosystem whose multiple interactions and self-reinforcing feedback loops are largely unknown and whose outcomes are therefore uncertain.

Because outcomes of a shared vision are uncertain, human activities encompassed by a vision can be thought of as tentative probing into various aspects of Nature and are best taken one step at a time and tested at each step. Through such testing, one hopes to detect potentially adverse and unpredicted effects at an early stage so that potentially deleterious activity can be corrected to the best of one's ability *before* serious, widespread, or irreversible damage occurs.

Because of the uncertainties of the future, the best monitoring and the best adjustments (target corrections) based on that monitoring provide several preplanned actions, such as: "If A happens, I will do B; if C happens, I will do D;" and so on. Otherwise, one monitors only outcomes and merely hopes necessary corrective actions will be found in time.

An example might be the assumption that placing fish ladders in dams will sustain the ability of salmon to migrate and thus perpetuate their survival. Yet over time, we have learned that the reservoirs created by the dams (in addition to a host of other human activities) also affect the survival and migration of the salmon. Accordingly, monitoring a single variable, even a seemingly reasonable one, such as counting fish, may do nothing to clarify the issue or save the fish, which was the original purpose of the fish ladders. This scenario points to the necessity of accepting and remembering that things are inevitably more complex than we are able to foresee, the very reason we need to monitor in the first place.

Good monitoring has seven steps: (1) crafting a vision and goals; (2) preliminary monitoring or inventory; (3) modeling our understanding, (4) writing a caretaking plan with clearly defined objectives, (5) monitoring implementation; (6) monitoring effectiveness; and (7) monitoring to validate the outcome(s).

Step 1: Crafting a vision and goals—Crafting a carefully worded vision and attendant goals that state clearly and concisely what you want as a future condition is the necessary first step in monitoring so that you know what you want, where you want to go, and what you think the journey will be like. The vision and its goals form the context of the journey against which you measure (monitor) all decisions, actions, and consequences to see if in fact your journey is even possible as you imagined it and what the consequences of the journey might be. Then, and only then, are you ready for the next steps in monitoring.

Step 2: Preliminary monitoring or inventory—Preliminary monitoring is to carefully observe and understand the circumstances that you begin with, taking "inventory" of what is available here, now. Taking inventory requires these questions: (1) What exists now, before anything is purposefully altered? (2) What condition is it in? and (3) What is its prognosis for the future? Even though preliminary monitoring may require multiple questions, the outcome is still a single realization.

If, for example, you go to your doctor for an annual checkup, your doctor would have to take a series of measurements, such as your blood pressure and blood chemistry, and would have to know what a healthy person is (including, if possible, you as a healthy person) as a benchmark against which to judge your current condition. If you are indeed healthy, then all is well; if not, your doctor would presumably prescribe tests to pinpoint what is wrong, prescribe medicine to correct your ailment, and make a prognosis for your future.

If you go to your doctor, but only allow your blood pressure to be taken without checking the level of your cholesterol, your doctor cannot deal with your health as a systemic whole and loses the ability to see the various components as parts of an interactive system. In this, your body is similar in principle and function to your family, the community in which your family resides, the landscape in which your community rests, and the landscape within the bioregion.

Step 3: Modeling collective understanding—By modeling, I mean configuring the current knowledge of the system of interest into a conceptual model as an explicit map of people's collective understanding. Such a map could be augmented by one or more computer models. Although it is critical in this exercise to assume at the outset that such a map represents our best understanding of the system, it is equally critical that we have the humility to assume the map is flawed and our understanding incomplete. In this way, the viability of the vision and its goals, as well as the model itself, are continually tested and improved. As the model is improved, so is our knowledge of the system with which we are working.

Step 4: Writing a caretaking plan — Each forest would have its own treatment plan (= "management plan" in the current parlance) in which all the particulars of its care, including objectives, would be laid out. An *objective* is a specific statement of intended accomplishment. It is attainable, has a reference to time, is observable and measurable, and has an associated cost. The following are additional attributes of an objective: (1) it starts with an action verb; (2) it specifies a single outcome or result to be accomplished; (3) it specifies a date by which the accomplishment is to be completed; (4) it is framed in positive terms; (5) it is as specific and quantitative as possible and thus lends itself to evaluation; (6) it specifies only *what, where, when, quantity*, and *duration* and avoids mentioning "why" and the "how;" and (7) it is product oriented.

Once you have determined your objective(s), you not only will be able to but also must answer the following questions concisely: (1) What do I want? (2) Where do I want it? (3) When do I want it? (4) How much (or how many) do I want? (5) For how long do I want it (or them)? If a component is missing, you may achieve your desire by default, but not by design.

A simple illustration would be to ask a linear-minded CEO of an exploitive timber company what his main objective is. If he was completely honest, his answer would be: "I want all the public's old-growth trees at my mill now, just long enough to cut them into lumber."

If we break his statement into its components as questions, it looks like this: (1) What do you want? "All the public's old-growth trees." (2) Where do you want them? "At my mill." (3) When do you want them? "Now." (4) How many do you want? "All of them." (5) How long do you want them for? "Just long enough to cut into lumber."

Only when you can answer all of these questions concisely do you know where you want to go and the value of going there, and only then can you calculate the probability of arrival. Next you must determine the cost, make the commitment to bear it, and then commit yourself to keeping your commitment.

Before a single treatment plan is written, however, the professional caretakers (from foresters, biologists, and engineers to secretaries, accountants, road engineers, and building custodians) would tour the area to be treated in order to see what it looks like, ask questions, and learn of perceived problems and potential, corrective actions. This tour would not only be a reality check for the objectives in terms of the vision and goals but also would help each person see the importance of their job in relation to the forest legacy that will greet the next generation. It would also pass the various people through the four gates of a trusteeship, which is important because the final forest legacy would, after all, be of their design as a team, and they must each be comfortable with it.

When possible, a class of grade-school children would also tour the site and tell the team what they—as the beneficiaries who must live with the outcome—want it to be like when they grow up and have children of their own. In this way, the beneficiaries would have a voice with the trustees, and the trustees would have the counsel of the beneficiaries.

There would also be a post-treatment tour for both the trustees and the beneficiaries in order to assess the outcome of the treatment, to see what did or did not work, discuss the ramifications and potential alternatives, and improve the next treatment. *Expensive!* you might say. True, but far less than the $1 million for each Tomahawk missile that swooped into an Afghani cave, the $145,000 for each "bunker-buster" bomb—which is about the same as the price of a median American home, and then there is the $13,700 an hour that is costs to fly each B-2 bomber. In addition, two helicopters crashed during the first two months of the Afghan war at a cost of $11 million and $40 million, respectively,[206] to say nothing of the ongoing costs in Afghanistan and the invasion of Iraq. The latter has cost Americans $135 billion as of September 30, 2004, according to the Library of Congress' database. Considering that the current administration has continually asked for—and been given—more money to destroy parts of the world abroad, I would think that we should at least be able to repair part of our world here at home for a much lesser cost. To me, expensive is relative, especially when the quality of life for all generations of Americans is at stake. Priorities are the real issue.

Step 5: Monitoring implementation—Monitoring the implementation of a project on the ground asks: Did we do what we said we were going to do?

While this type of monitoring is really just documentation of what was done, it is critical documentation because without it, it may not be possible to figure out what went awry (if anything did or does), how or why it went awry, or how to remedy it. In addition, the next generation would have little or no idea of what you (as forest caretaker) did or why and thus no way to figure out what to do in order to remedy a problem that has arisen since you either retired and moved away or died.

To continue the doctor analogy, it is important to document whether your doctor really did the test deemed medically necessary. Certainty about a given test is crucial because, if there is doubt whether a certain test was performed or how well it was performed, the dubious outcome of one test can seriously compromise the results of another test and so might completely alter (perhaps disastrously) the doctor's interpretation of the test's outcome.

Step 6: Monitoring effectiveness—Monitoring to assess effectiveness means monitoring to assess the implementation of your objectives, not the goals or vision. A vision and its attendant goals describe the desired future condition for which you are aiming. They are qualitative and not designed to be quantified. An objective, on the other hand, is quantitative and so is specifically designed to be quantifiable.

Monitoring to assess effectiveness of an objective requires asking: Is the objective specific enough? Are the results clearly quantifiable and within specified scales of time? Systematically monitoring the effectiveness of your project with the aid of indicators provides information (feedback) that allows you to assess whether you are in fact headed toward the attainment of your collective vision, maintaining your current condition, or moving away from your collective vision.

A good indicator helps a forest caretaker recognize potential problems and provides insight into possible solutions. What a caretaker chooses to measure, how they choose to measure it, and how they choose to interpret the outcome would have a tremendous effect on the biological viability of the forest in the long term.

Indicators close the circle of action by both allowing and demanding that you come back to your beginning premise and ask (reflect on) whether, through your actions, you are better off now than when you started: if so, how; if not, why not; if not, can the situation be remedied; if so, how; if not, why not; etc., etc.

Here a caution is necessary. Traditional unidimensional indicators that measure the apparent health of one condition (say the trees in a single stand), ignore the complex relationships among soil, water, air quality, and the relationship of neighboring stands within the water-catchment. When each component is viewed as a separate issue and monitored in isolation, measurements tend to become skewed and lead to ineffective policies that in turn can lead to a deteriorating quality of the forest. This being the case, indicators must be multidimensional and must measure the quality of relationships among the components of the system being monitored if the forest in question is to have any kind of accurate assessment of its sustainable well-being.

Only with relevant indicators and a systematic way of tracking them is it possible to make a prognosis for the future based on your vision and goals (which state the desired condition) and the collective objectives (which determine how the goals will be achieved and when). Only with relevant indicators and a systematic way of tracking them is it possible to make the necessary target corrections to achieve the vision and its attendant goals because only now can you know what corrections to make.

Returning to the doctor analogy, assessing effectiveness asks: Was the right test used (was it relevant to your condition)? Was the test effective (if it was the right test, did it perform as it was supposed to)? These are important questions, because if the wrong test was used or the test was ineffective, the results can be very different from those you were led to expect and the outcome could be unexpectedly life threatening.

Step 7: Monitoring to validate the outcome—Monitoring for validation of the outcome is considered by many to be research. This type of monitoring involves testing the assumptions that went into the development of the objectives and the models they are based on.

Monitoring for validation may require asking such questions as: Why didn't the results come out as expected? What does this mean with respect to our conceptual model of how we think the system works versus how the system actually works? Will altering our approach make any difference in the outcome? If not, why not? If so, how and why? What target corrections do we need to make in order to bring our model in line with how the system really works?

Validation is a necessary component of any monitoring plan because this is where you learn about the array of possible target corrections. In addition, monitoring for validation may have wide application for other projects.

Visiting once again the doctor analogy, suppose your doctor says: "The results of your tests show that you may have an unusual form of 'Dingy Disease.' Let me check with my colleagues and the literature to see what is known about it."

Three weeks later, your doctor calls you into the office and says: "Your particular form of 'Dingy Disease' is indeed unusual, and I can't find a common cure. Luckily, there's a drug called 'Dumpin Dingy' that shows promise in laboratory tests. It's now ready for human trials to see how well they corroborate laboratory results. If you're willing to become part of a controlled medical experiment, I think you may qualify for the drug. It's your best chance at the moment to regain your health."

You take part in the drug experiment because you are assured that it will be carefully monitored under strictly controlled conditions, and you have faith in what you are told.

Here, you must keep in mind that dealing with a medical issue on a personal basis is simpler in many ways (except perhaps to the individual and immediate family and friends) than dealing with the multiple, ecological variables of land, over which we humans have no control. Regardless, both cases, personal and ecological, have unknowable outcomes that are projected into an unknowable future.

While it is necessary in today's world for professional caretakers to focus on and be responsible for a single vision of the collective, public forests as a biological living trust, the actual care of the various parts of the forest may be administered by two or more jurisdictions, such as the Forest Service, the Bureau of Land Management, National Park Service, and the Fish and Wildlife Service, within the same bioregion. This arrangement *can succeed*, despite the mix of jurisdictions, if people choose to make it work. Yet, even with the best intentions, there is a weakness in monitoring that must be overcome if the results are to really serve the purpose for which they were designed.

The Outcome

One of the major faults associated with the implementation of a shared vision is that it's based on short-term or high-frequency responses from the environment, whereas the effects of human behavior ripple through the invisible present into the distant future, often beyond the ability of short-term corrective actions to be of value. In the mere instant of short-term monitoring, I find no viable answer to Garrett Hardin's question: "And then what?"[207]

Surprises (usually unwelcome ones) are inevitable, and their seeds may or may not be entrained in the stream of data from monitoring and may or may not be discovered by those who observe the data because most monitoring programs are scaled to the immediate future and thus are poorly scaled to the slower, longer-term responses of the environment that are not immediately apparent due to the aforementioned "invisible present." The challenge in caretaking public forests is to see the correlation between and among information that occurs frequently and events at the next magnitude of scale that occur less frequently than the information from monitoring would indicate.

This is not to say that monitoring for short-term events is unimportant. It is only to point out that monitoring only for the short term will likely prove to be a disastrously myopic choice in ecosystems with long-term feedback loops and/or high degrees of environmental variability. Meanwhile, monitoring for the long-term, ecological effects of human activities usually fails—if it gets started at all—because humans have a tragically short attention span, one that leads to broken feedback loops of information, especially information that helps a caretaker decide whether to change the direction or intensity of a particular activity.

In the end, it is through the questions we ask that we derive our vision, goals, and objectives. It is the questions we ask that frame our perceptions and direct our actions, those that require monitoring. And it is the questions we ask that determine what and how we monitor. Finally, how we caretake our public forests depends first and foremost on the relevance of the questions we ask, for they become our compass and map into the future.

In Conclusion

A species with no commercial value, bristlecones have prospered unknown and undisturbed for millennia. It is precisely because they have been left alone that they have become so important. Bristlecone pines are an eloquent argument for non-management—for respecting what we don't know about a place and its ecosystems.

— Jane Braxton Little

Poet Matthew Arnold observed almost a century and a half ago that we live "wandering between two worlds, one dead, the other powerless to be born." Arnold could see that our human perception of Nature, reflected in our basic institutions, including science, was inconsistent with the world around us. I concur with Arnold's astute observation because a forest is cyclic, not linear, as most economists and most industrialists would have us believe. Therefore, to hold a discourse about any one piece is to hold a discourse about the many, a milieu of pieces, and their interconnectedness that quickly becomes incredibly complicated, especially since this milieu includes our social lack of psychological maturity.

Even today, with all our scientific knowledge, the new world is still unable to be born because all new experience demands letting go of the old and the dead—demands risking the living unknown, something we're loath to do. Instead, we insist on facing the unknown future of our rapidly changing world with concepts, methods, and institutions that remain tenaciously rooted in the old, comfortable experience.

We need to learn to see the forest as it is: a dynamic living organism that requires healthy soils to grow life among the trees and filter water; that produces pure water to drink and with which to irrigate crops; that produces habitat for the rearing of salmon and steelhead, as well as deer, elk, and a myriad other life forms; that produces the countless other products, amenities, and free services of Nature that are critical to the social stability and ecological sustainability of our communities; and that nurtures us humans as an inseparable, natural part of itself.

Even so, there is a prevailing, widespread notion afoot today in Western, industrialized nations that Nature is merely a commodity to be economically exploited. We are here, according to Judeo-Christian religion, to master Nature, and as masters, to improve Nature's ability to function in producing goods and services strictly for our benefit.[208] There is also the antithetical view that considers human beings as an artifact of Creation, as totally separate from Nature and so an unnatural intrusion into the world of Nature, where our very presence is defiling.

Regardless of what we have gleaned from our political sanctification by organized religions, we humans animal are still animals. We may not like being called animals or being thought of as animals, because we all too often look down on nonhumans. Nevertheless, we are animals.

What sets us apart from the rest of our fellow creatures is not some higher sense of spirituality or some nobler sense of purpose, but rather that we deem ourselves wise in our own eyes. And therein lies the fallacy of humanity. We are no better than or worse than other kinds of animals; we simply are a different kind of animal—one among the many. So it is that we are an *inseparable* part of Nature, not a special case apart from Nature.

Consider that "natural," as defined in today's dictionaries, means to be present in or produced by Nature, such as animals, as opposed to being made by humans, which is seen as artificial, an artifact. Consider further that we humans, by our very nature, are produced in a natural way, through sexual union, the joining of sperm and egg, and live birth; that human mothers produce milk in a natural way to nourish their offspring; that we grow in a natural way, through the division of our bodily cells; that we both eat and void our bodily wastes in a natural way; that we both awake and sleep in a natural way; that we sexually reproduce in a natural way; that we die in a natural way; and that our bodies, if left alone, decompose in a natural way. We are, therefore, natural by virtue of our creation. And to be natural is to be an inseparable part of Nature, from that which we are created.

As a part of Nature, is not what we do natural? This is not to say that our actions are wise, or ethical, or moral, or desirable, or even socially acceptable and within the bounds of Nature's inviolate laws. This is only to say that our actions are natural, because we are natural, and that includes converting the landscape from Nature's design to human society's, cultural design.

Admittedly, Nature's design may be more pleasing to our senses than are many of our cultural designs and may function in a more ecologically healthy way over time. Be that as it may, one design is as natural as the other—only different in context. Nevertheless, our cultural alterations are often fraught with a schism between spirituality and materialism, a schism that is knocking at the door of our collective consciousness.

Although I detect a definite psychological shift on the part of many people who are trying to reach back into human history, back into consciousness to accommodate Nature, to recapture some primordial sense of spiritual harmony with Nature, I still see most of us struggling with a sense of materialistic separateness—the context of our existence. Put another way, because we humans are a natural, inseparable part of Nature, an area "managed" as wilderness is just as natural as an area of forest that has been clear-cut, despite the fact that a clear-cut *emulates nothing in Nature*—not even fire. Nonetheless, a wilderness and a clear-cut differ only in context, a point often missed.

I find, when I speak to various groups throughout the United States and Canada, that audiences almost inevitably accept the area of designated wilderness as natural but not the area of forest that has been clear-cut. Yet both have been created by human society: a wilderness by the invention of a human law called the "Wilderness Act" that helps to feed humanity's need for spiritual renewal, and a clear-cut by the invention of a human idea and a human tool — capitalism and the chainsaw, both of which help to feed the timber industry's short-term, monetary motives.

Most people seem to know what happens to an area of forest that is clear-cut — a road is built; all the trees are cut down and removed; seedling trees may or may not be planted as a crop; and fire, unwanted vegetation, and insects are controlled as much as possible within the area. Oddly, these same people do not seem to know that it is *legal*, in the United States at least, to: mine in a wilderness area; graze it with livestock; hunt in it; trap fur-bearing mammals in it; fish in it; cut trails through it; cut wood in it for camp fires; trample it with horses, llama, and human feet; and at times even control fire and epidemic outbreaks of insects in it. And in the long run, we can and do have a greater, cumulative effect of negative outcomes on some areas of wilderness than we do in some areas that have been clear-cut.

Both a wilderness and a clear-cut are natural. The difference between them is only a matter of the perceived motive governing the type and the degree of human manipulation. In other words, the difference between them is the perceived degree of "naturalness" based on some context of personal acceptability with respect to the observable human disturbances of the environment.

And yet, because we still see ourselves as somehow separate from Nature, the connotation of "natural" is of something apart from human society, a purity without contamination by human activity or artifact. Clearly, this connotation of "natural" is invalid because humanity and human society are inseparable from both Nature and the natural world — especially forests.

To maintain healthy forests, we need to learn about reinvesting biological capital into their ecological processes so our mills will have a sustainable yield and harvest of timber. We need to understand that Nature cannot be constrained to absolutes, that a sustainable yield is a trend within some limits, that even the timber industry must be flexible and continually change as the forest changes. And our schools of forestry must become leaders in research, caretaking, real innovation, and human relations rather than the last bastions against inevitable change.

In addition, the timber industry steadfastly resists changing its exploitive practices at the expense of our global environment. I say this because more than any other enterprise of humanity, the timber industry has, and is, negatively affecting the forests of the world, without considering, let alone cataloging, the deleterious affects on all forest-dependent industries worldwide.

Yet still today, the driving economic force in our nation's forests is the extractive timber industry that the public perceives as *the* "forest industry," an industry that goes primarily from the forest to the mill. How have we let it happen that the timber industry has for so long been the only forest-dependent industry with a voice? We are in essence allowing this one interest to speak for all forest-dependent industries while in truth it's looking out solely for its own self-interest by grossly altering the ecological sustainability of the world's forests for short-term economic profits. In addition, the timber industry is continuing to eliminate many jobs by shipping logs to other countries for processing, by automating its mills to get rid of people, and by hostile takeovers of smaller timber companies, all justified in the euphemistic terms of "economic competition."

Clearly, in our linear, product-oriented thinking, an old-growth forest is an economic waste if its "conversion potential" is not realized—that is, the only value the old-growth trees have is their potential for being converted into money. Conversion potential of resources counts so heavily because the effective horizon in most economic planning is only five years away. Hence, in our traditional, linear, economic thinking, any merchantable old tree that falls and reinvests its biological capital into the soil is an "economic waste" because its economic potential was not immediately converted into monetary profit.

With both conversion potential and economic competition in mind, new equipment is constantly being devised to make harvesting timber ever more *efficient*. The chainsaw, for example, greatly speeded the liquidation of old-growth forests worldwide. Possessed by this new tool, most folks in the timber industry and the forestry profession lost all sense of restraint and began cutting forests faster than they could re-grow. Further, no forested ecosystem has yet evolved to cope ecologically with the massive, systematic, and continuous clear-cutting made possible by the chainsaw and the purely economic motivation behind it.

As the timber industry affects the ecological sustainability of the forest for short-term profits, it has a dramatic effect on *all* forest-dependent industries—often to the detriment of the long-term product base of those industries. For Oregon, the Pacific Northwest, and indeed the world, we must have a balance between the short-term profits for the timber industry and the long-term, economic sustainability of all other forest-dependent industries. Whatever balance is struck, it must, of social necessity, favor first and foremost the ecological sustainability of the forest, which produces the water that sustains all life.

It is clear that the timber industry must be redefined, redesigned, and restructured. It's equally clear that such change must render the timber industry dramatically different than it is today. After all, it is only *one* of a suite of the interdependent, forest-dependent industries that serve the economic necessities of humanity. This being the case, foresters of the future will need to:

- understand the parts of a forest in relation to its organic whole and the whole in terms of its parts by learning to sit with and in the forest and hear its voice.
- understand the science of each component and subsystem of the forest,

i.e., to be able to evaluate and openly question new information in terms of the old, old information in terms of the new, and both in terms of existing ignorance;

- understand enough about the ecological linkages among forest components and subsystems to be able to anticipate effects of management stresses;

- understand and work with a forest as a dynamic component of an ever-evolving, culturally oriented landscape while simultaneously honoring the integrity of the blueprint of Nature's processes and patterns;

- understand that the only sustainability for which we can "manage" is that which ensures the ability of an ecosystem to adapt to evolutionary change, which means managing for choice (maximum biodiversity);

- understand how to manage for a desired condition of the landscape and abandon the unworkable notion of sustained ever-increasing yield of natural resources;

- understand how to manage for the connectivity of habitats to help ensure the ecological wholeness and the biological richness (biodiversity) of the patterns they create across the landscape;

- be able to abstract, simplify, synthesize, and generalize information about complex systems so that his or her "intuitive mind" can act as the final reality check of relevant information prior to making decisions;

- be able to articulate ideas effectively, clearly, and accurately both in writing and in public speaking; and

- be able to work openly and skillfully with people with sufficient knowledge and in sufficient depth to validate their concerns and to give them the critical understanding and trust of the professional rationale behind a decision, even if they are opposed to it.

Today, the sustainability of our public forests, as our biological living trust, rests—as never before—in our hands and in the wisdom of our decisions that, for better or worse, will determine our legacy for tomorrow, a legacy that's becoming increasingly irreversible. By way of illustration, consider a violin as a symbol of the fine-grained wood from our ancient forests. Then consider such composers as Johann Sebastian Bach, Wolfgang Amadeus Mozart, Ludwig van Beethoven, and Franz Schubert, each of whom committed to paper music of great beauty, the translation of which for more than a century has been through orchestras. An orchestra, in turn, is composed of musical instruments and musicians that together give "voice" to the mute notes on paper. And a musician's ability to play an instrument is dependent not only on human skill but also on the quality of the instrument.

Over the last two centuries, the violins made by Antonio Stradivari have given the human ear some of the world's most exquisite melodies. Indeed, the wood of the ancient trees and the loving labor with which Stradivari crafted the wood to perfect his violins is a marriage of such harmony that each violin he made is called a "Stradivarius."

I've heard it said that a bird does not sing because it has an answer; it sings because it has a song. In the same sense, musicians do not play because they have an answer; musicians play because they have a melody and a fine instrument, such as a Stradivarius, with which to give "voice" to that melody.

To build a violin with the quality of a Stradivarius, one must be an expert violinmaker and also have available fine-grained wood from ancient trees. The quality of a Stradivarius is of yesterday and of today, but what about the violins of tomorrow? I was once confronted with this question in Seattle, Washington, where I spoke at a conference.

A young man who made violins entirely by hand came to me and asked: "What will I do for a living when the ancient forests are all gone and there's no more special, high-quality wood, such as that of an ancient, tight-ringed, clear-grained Sitka spruce, for me to work with?" He paused for a moment and then continued, "There are very few people who work the way I do, and we're rapidly becoming fewer."

Think carefully about his question. If we continue to convert sustainable forests into nonsustainable tree plantations that, among other things, deplete soil fertility and produce inferior-quality wood, the symphonic beauty of the centuries will become a hollow echo of dull tones as they play a requiem for our forests. This outcome is not necessary. To forestall it, however, we, in the United States, must learn to slow down and accept a more leisurely way of living, as well as one that is economically and ecologically sustainable.

The Chinese character for leisure is composed of two elements that, by themselves, mean "open space" and "sunshine." Hence, an attitude of leisure creates an opening that allows the sunshine in. Conversely, the Chinese character for busy is also composed of two elements that, by themselves, mean "heart" and "killing." This character points out that for the beat of one's heart to be healthy, it must be leisurely, an attribute we tend to think of as a privilege reserved for the well-to-do.

We, in our hurry, worry, action-product-oriented world of today, seem never to have the time to do something right the first time, but we seem also to have time to do it over and over and over again. Do we really lack the inherent ability to simplify and cull the inconsequential from the basic necessities? The answer is a resounding, "No."

". . . leisure," according to Brother David Steindl-Rast (a Benedictine Monk), "is a virtue, not a luxury. Leisure is the virtue of those who take their time in order to give to each task as much time as it deserves. . . . Giving and taking, play and work, meaning and purpose are perfectly balanced in leisure. We learn to live fully in the measure in which we learn to live leisurely,"[209] a sentiment echoed by Henry David Thoreau: "The really efficient laborer will be found not to crowd his day with work, but will saunter to his task surrounded by a wide halo of ease and leisure." Might this become the new American work ethic?

Because I accept the foregoing postulate, it is my contention that leisure is a prerequisite for the proper caretaking of the public forests. To this end, it would be good business to arrange workshops that help people understand the true meaning and value of leisure and to create an atmosphere that honors leisure in the caretaking of our public forests. Here, it must be understood that self-mastery is part of leisure.

"Someday, after we have mastered the winds, the waves, the tide and gravity, we shall harness for God the energies of love. Then, for the second time in the history of the world, man will have discovered fire." So wrote Teilhard de Chardin, the noted French Jesuit priest, paleontologist, and philosopher.

As we learn to master ourselves and our own creative processes, we have the potential to harness and enlist the energies of love in helping us become the dominant creative force in our own lives and in the lives of others, present and future, especially where public lands are concerned. As individual adults and leaders, let us reconnect ourselves to the passion of our highest ideals and potential, our deepest self-knowledge, and our own inner truth because each generation must be the conscious keeper of the generation to come—not its judge.

As the keeper of the next generation, it is our duty to prepare the way for those who must follow. This entails, among other things, wise and prudent caretaking of our public lands in order to safeguard the very best of our human values by carefully, purposefully interweaving them into the fabric of our being and our social relationships in space (with one another and the land) and time (among generations). With this in mind, let's once again rememberit the ecological principles that form the basis of caretaking the public forests.

We *can* have sustainable forests, but only if that is what we are committed to and only if we act now. We can have sustainable forests, but only if we constantly question and re-evaluate along the way what we think we know and only if we retain all of the pieces—including indigenous forests in all successional stages (legacy forests)—to learn from. We can have sustainable forests, but only if processes-oriented systems-thinking and product-oriented, linear-minded thinking are complementary in a way that first and foremost protects the biophysical sustainability of our national forests—both public and private. We can have a sustainable suite of forest-dependent industries, including the timber industry, to produce products for people, but only if we redefine and redesign the timber industry to operate, in fact, within the ecological limitations set by the forest, not by linear-minded economists, accountants, shareholders, and corporate-political board members.

In all cases, we must:

- learn humility, meaning we must learn to be teachable by Nature
- become students of processes—not advocates of positions
- grow beyond the limited view of our own narrow interests
- work together for a common goal, with a common commitment: a sustainable forest for a sustainable environment for a sustainable suite of forest-dependent industries for sustainable human communities, for a sustainable quality of life—present *and* future

But before we can maintain or restore sustainable forests, we must have the humility to sit in Nature's classroom—the indigenous forest—with a beginner's mind, a mind that is simply open to the wonders of the present and the positive possibilities of the future. Here, away from corporate-political boardrooms, our scientific knowledge can blend with our intuition. Here, with pen in hand and an inkwell called choice, we can craft a visionary plan that protects both the sustainability of our public forests as a biological living trust and helps to maintain the economic sustainability of the forest-dependent industries upon which the social well-being of our communities depends—our legacy, our bequest to the beneficiaries, the children of all generations.

If we are to caretake our public forests in such a way as to create a "wise person's forest," one that will meet the changing needs of successive generations, we must ask a vital question: Do we owe anything to the future? If so, we must understand and accept there are no external fixes for internal, moral imperatives; there are only internal shifts of consciousness and morally correct intentions and behaviors. We must also understand and accept that all we can bequeath the generations of the present and of the future are options—choices to be made, the right to choose as we have done, and some things of value from which to choose.

To protect that right of choice, we must grant all children the right of free speech and the right to be heard. We must ask new, responsible, relevant, future-oriented questions, questions that determine the quality of lifestyle we wish to have and that we wish our children to be able to have. With answers in hand, we need to determine first and foremost how much of a given resource is necessary to leave intact in the environment as a biological reinvestment in the health and continued productivity of the ecosystem. We must, at any cost, be it economic or political, protect the quality of the soil, water, and air of our home planet if humanity and its society is to survive. We must also view the environment from the standpoint of biological and cultural necessities as opposed to limitless personal wants, desires, needs, and demands, and, if necessary, alter our lifestyles to reflect what the ecosystem can in fact sustainably support.

We must account for the intrinsic, ecological value of all natural resources— *not just* their conversion potential into money, and we must accept that the long-term health of the environment takes precedence over the short-term profits to be made through nonsustainable, exploitive development. Concurrently, we must convert our society (immediately, consciously, and unconditionally) to a version of capitalism that views long-term ecological wholeness and biological richness of the environment as *the measure* of long-term economic health.

To this end, it is imperative that we pass clearly stated, precisely worded, unambiguous laws, wherein the intent is so simply stated that it cannot be distorted and hidden by internal bureaucratic policy that is either inept or conceived to purposefully circumvent a given law or legal mandate. We must create social-environmental policy that is commensurate with social-environmental sustainability, while concurrently promulgating policy that protects the ecological integrity of the environment from the negative, irreversible aspects of nonsustainable development. With respect to our public forests, such policies are perhaps best framed and articulated by the Forest Advisory Council (with advice from the Children's Council) and then put to a popular vote of all citizens to protect them from the self-serving agendas of the agencies at the mercy of self-serving agendas of *some* members of Congress, *of some* Presidential Administrations, and *some* corporations and companies.

We also must accept that the only sustainability for which we can caretake an ecosystem, such as a public forest, is one that ensures the ability of the ecosystem to adapt to changing environmental conditions in a way that is at least benign with respect to our human existence. This means we must caretake for choice—ecological diversity in *all* its various forms, regardless of the economic and political costs. In turn, ecological diversity can be protected only by caretaking an ecosystem within the constraints of a clearly crafted vision for social-environmental sustainability and by *abandoning* our cherished, unworkable notion of a *sustained, ever-increasing yield of industrial resources*. To achieve such a desired condition, we must caretake for the spatial and temporal connectivity of habitats and stop today's practice of "managing" *for* fragmentation of the landscape by focusing only on commodity-producing resources. I say this because the connectivity of habitats — as the bedrock of caretaking our forests — will help to ensure the ecological wholeness and biological richness of the patterns we create across the landscape, even as we extract resources for our use.

If we are to be successful trustees of the future's right of choice, we must unfailingly *manage* the *only thing* we really can manage—*ourselves*, as well as what *we introduce* into the environment. Now is the hour to act. This is the decade that will set the course by which humanity will navigate the 21st century. Ours is perhaps the greatest human challenge in history — to become a nation of psychologically mature adults and lead the world into an era of social-environmental sustainability.

We must begin to care for our public forests as a biological living trust. I know that as individual caretakers, our accomplishments are limited, but together, we can become truly awesome as trustees for the children and the environmental health of the Earth they will inherit. Ours is the legacy. Theirs is the consequence. What kind of legacy shall we choose?

Endnotes

1. The preceding two lists of tree characteristics are modified from: Richard H. Waring and Jerry F. Franklin. "Evergreen coniferous forest of the Pacific Northwest." *Science* 204 (1979):1380-1386.

2. Chris Maser. *Mammals of the Pacific Northwest: From the Coast to the High Cascade Mountains.* Oregon State University Press, Corvallis, OR. 1998. 406 pp.

3. The preceding discussion of decomposing wood is based on: (1) Chris Maser and James M. Trappe (Technical Editors.) *The Seen and Unseen World of the Fallen Tree.* USDA Forest Service General Technical Report PNW-164. U.S. Department of Agriculture, Forest Service, Pacific Northwest Forest and Range Experiment Station, Portland, OR. 1984. 56 pp; (2) Chris Maser, James M. Trappe, and C.Y. Li. "Large Woody Debris and Long-Term Forest Productivity." *In: Proceedings of the Pacific Northwest Bioenergy System: Policies and Applications.* Bonneville Power Administration, May 10 and 11, Portland, OR. 1984. 6 pp; (3) Chris Maser and James M. Trappe. "The Fallen Tree—A Source of Diversity." Pp. 335-339. *In:* Forests for a Changing World. *Proceedings of the Society of American Foresters 1983 National Conference.* 1984; (4) Chris Maser, Robert F. Tarrant, James M. Trappe, and Jerry F. Franklin (Technical Editors). *From the forest to the sea: A story of fallen trees.* USDA Forest Service General Technical Report PNW-229. Pacific Northwest Research Station, Portland, OR. 1988. 153 pp; and (5) Chris Maser. *Ecological Diversity in Sustainable Development: The Vital and Forgotten Dimension.* Lewis Publishers, Boca Raton, FL. 1999. 401 pp.

4. Chris Maser. *Forest Primeval: The Natural History of an Ancient Forest.* Oregon State University Press, Corvallis, OR, 1989. 282 pp.

5. Rutherford Platt. *The Great American Forest.* Prentice-Hall, Inc., Englewood Cliffs, N.J. 1965. 271 pp.

6. Aldo Leopold. *A Sand County Almanac.* Oxford University Press, NY. 1966. 269 pp.

7. Clyde S. Martin. "Forest resources, cutting practices, and utilization problems in the pine region of the Pacific Northwest." *Journal of Forestry* 38 (1940):681-685.

8. Greenpeace. *Partners in Mahogany Crime: Amazon at the Mercy of 'gentlemen's agreements.'* Greenpeace International, Amsterdam, The Netherlands. 2001. 18 pp.

9. (1) Chris Maser. *The Perpetual Consequences of Fear and Violence: Rethinking the Future.* Maisonneuve Press, Washington, D.C. 2004. 388 pp. and (2) John Gowdy (editor). *Limited Wants, Unlimited Means.* Island Press, Washington, D.C. 1998. 342 pp.

10. If you want to know what I mean by "social-environmental sustainability," see: Chris Maser. *Sustainable Community Development: Principles and Concepts.* St. Lucie Press, Del Ray Beach, FL, 1997. 257pp .

11. Nate Wilson. "Not so Peachy Forestry at U of Ga." *Distant Thunder, Journal of the Forest Stewards Guild,* 11 (2001):10-11.

12. Lester Brown. "A Copernican Shift." *Resurgence* 213 (2002):14-15.

13. The Associated Press. "Washington to launch new master's program." *Albany (OR) Democrat-Herald, Corvallis (OR) Gazette-Times.* January 24, 1999.

14. The preceding discussion of the "soil-rent theory" is based on: Richard Plochmann. "Forestry in the Federal Republic of Germany." *Hill Family Foundation Series,* School of Forestry, Oregon State University, Corvallis, OR. 1968. 52 pp.

15. The preceding four paragraphs are based on: P. Bak and K. Chen. "Self-organizing criticality." *Scientific American.* January(1991):46-53.

16. The preceding discussion of carbon dioxide is based on: (1) Kathleen C. Weathers, Gene E. Likens, F. Herbert Bormann, and others. "Cloudwater Chemistry from Ten Sites in North America." *Environmental Science & Technology* 22 (1988):1081-1026; (2) William H. Schlesinger. *Biogeochemistry: An Analysis of Global Change*, 2nd ed. Academic Press, San Diego, CA. 1997. 588 pp; and (3).Rebecca M. Shaw, Erika S. Zavaleta, Nona R. Chiariello, and others. "More CO_2 Lowers Plant Productivity." *Science* 298 (2002):1987-1990.

17. (1) George H. Taylor and R.R. Hatton. "The Oregon Weather Book." Oregon State University Press, Corvallis, Oregon. 1999. 240 pp.; (2) George H. Taylor and C. Hannan. "The Climate of Oregon." Oregon State University Press, Corvallis, Oregon. 1999. 212 pp.; and (3) N. J. Mantua, S. R. Hare, Y. Zhang, and others. "A Pacific interdecadal climate oscillation with impacts on salmon production." *Bulletin of the American Meteorological Society*, 78(1997): 1069-1079.

18. (1) Michael P. Amaranthus, Debbie Page-Dumroese, Al Harvey, and others. *Soil Compaction and Organic Matter Affect Conifer Seedling Nonmycorrhizal and Ectomycorrhizal Root Tip Abundance and Diversity.* USDA Forest Service Research Paper PNW-RP-494. U.S. Department of Agriculture, Forest Service, Pacific Northwest Research Station, Portland, OR. 1996. 12 pp.; (2) Robert Fogel. *Roots as primary producers in below-ground ecosystems.* Pp. 23-26. *In:* A.H. Fitter, D. Atkinson, D.A. Read, and M.B. Usher (eds.). "Ecological interactions in soil: plants, microbes, and animals." Blackwell Science Publications, Palo Alto, CA. 1985; (3) Daniel L. Luoma and Efren Cazares. *Effects of Ripping Compacted Soils on the Diversity and Abundance of Ectomycorrhizal Fruiting Bodies.* Final Report PNW-95-0731. Department of Forest Science, Oregon State University, Corvallis, OR. 2000. 10 pp., and (4) Gary M. Lovett, Kathleen C. Weathers, and Willian V. Sobczak. "Nitrogen saturation and retention in forested watersheds of the Catskill Mountains, New York." *Ecological Applications* 10 (2000): 73-84; and (5) Bernard T. Bormann. "Is There a Social Bias for Biological Measures of Ecosystem Sustainability?" *Natural Resource News* 3 (1993): 1-2.

19. H. Josef Herbert. "U.S. eases clean air regulations." *The Associated Press.* In: *Corvallis Gazette-Times*, Corvallis, OR. January 1, 2003.

20. Maja-Lena Brämvall, Richard Bindler, Ingemar Renberg, and others. "The Medieval Metal Industry Was the Cradle of Modern Large-Scale Atmospheric Lead Pollution in Northern Europe." *Environmental Science & Technology* 33 (1999): 4391-4395.

21. (1) Robert Denison, Bruce Caldwell, Bernard Bormann, and others. "The Effects of Acid Rain on Nitrogen Fixation in Western Washington Coniferous Forests." *Water, Air, and Soil Pollution* 8 (1977): 21-34; (2) Mark E. Fenn, Mark A. Poth, John D. Aber, and others. "Nitrogen Excess in North American Ecosystems: Predisposing Factors, Ecosystem Responses, and Management Strategies." *Ecological Applications* 8 (1998): 706-733; (3) Albert Tietema, Claus Beier, Pieter H.B. de Visser, and others. "Nitrate leaching in coniferous forest ecosystems: The European field-scale manipulation experiments NITREX (nitrogen saturation experiments) and EXMAN (experimental manipulation of forest ecosystems)." *Global Biogeochemical Cycles* 11 (1997): 617-626; (4) Steven G. McNulty, John D. Aber, and Steven D. Newman. "Nitrogen saturation in a high elevation New England spruce-fir stand." *Forest Ecology and Management* 84 (1996): 109-121; (5) Willian T. Peterjohn, Cassie J. Foster, Martin J. Christ, and Mary B. Adams. "Patterns of nitrogen availability within a forested watershed exhibiting symptoms of nitrogen saturation." *Forest Ecology and Management* 119 (1999): 247-257; and (6) J. Nilsson. *Critical Loads for Nitrogen and Sulphur.* Miljøapport 11, Nordic Council of Ministers, Copenhagen. 1968. 232 pp.

22. The preceding discussion of the six flaws in the "soil-rent theory" is based on:

Chris Maser. *Sustainable Forestry: Philosophy, Science, and Economics*. St. Lucie Press, Delray Beach, FL. 1994. 373 pp.

23. U.S. Laws, Statutes, etc. Public Law 86-517. [H. R. 10572], June 12, 1960. An act to authorize and direct that the National Forests be managed under principles of multiple use and to produce a sustained yield of products and services, and for other purposes. *In its* United States statutes at large. 1960. Vol. 74, p. 215. U.S. Government Printing Office, Washington, D.C. 1961. [16 U.S.C.. sec. 528-531 (1976).]

24. Richard Plochmann. *The Forests of Central Europe: A Changing View*. Pp. 1-9. *In*: Oregon's Forestry Outlook: An Uncertain Future. The Starker Lectures. Forestry Research Laboratory, College of Forestry, Oregon State University, Corvallis, OR. 1989.

25. The preceding two paragraphs are based on: Chris Maser, Russ Beaton, and Kevin Smith. *Setting the Stage for Sustainability: A Citizen's Handbook*. Lewis Publishers, Boca Raton, FL. 1998. 275 pp.

26. Martha Mendoza. "Stealing the forest for the trees." *The Associated Press*. In: *Albany (OR) Democrat-Herald, Corvallis (OR) Gazette-Times*. May 18, 2003.

27. The preceding discussion of declining fish stocks is based on: (1) Steve Newman. "Declining Fish Stocks." *The Los Angeles Times Syndicate*. In: *Albany (OR) Democrat-Herald, Corvallis (OR) Gazette-Times*. February 24, 2002; (2) The Associated Press. "Fish decision worries some communities." *Corvallis Gazette-Times*, Corvallis, OR. June 19, 2002; and (3) Ransom A. Myers and Boris Worm. "Rapid worldwide depletion of predatory fish communities." *Nature* 423 (2003): 280–283.

28. Edward Hyams. *Soil and Civilization*. Thames and Hudson, New York, NY. 1952. 312 pp.

29. John Perlin. *A Forest Journey: The Role of Wood in the Development of Civilization*. W.W. Norton and Co., New York, NY. 1989. 445 pp.

30. The preceding discussion of resource exploitation is based on: (1) Donald Ludwig, Ray Hilborn, and Carl Walters. "Uncertainty, resource exploitation, and conservation: lessons from history." *Science* 260 (1993): 17, 36; (2) Chris Maser, Russ Beaton, and Kevin Smith. *Setting the Stage for Sustainability: A Citizen's Handbook*. Lewis Publishers, Boca Raton, FL. 1998. 275 pp; (3) Russ Beaton and Chris Maser. *Reuniting Economy and Ecology in Sustainable Development*. Lewis Publishers, Boca Raton, FL. 1999. 108 pp.; (4) Scott Sonner. "Government wants equal footing." *The Associated Press*. In: *Albany (OR) Democrat-Herald, Corvallis (OR) Gazette-Times*. May 18, 2003; (5) *The Associated Press*. "Bush proposes 'charter forests' to give locals more control." *Corvallis Gazette-Times*, Corvallis, OR. February 6, 2002; and (6) Tom Kenworthy. "Proposal would ease way for roads in wilds." *USA Today*. March 6, 2002.

31. James Allen. *As a Man Thinketh*. Grosset & Dunlap, New York, NY. 1981. 72 pp.

32. Aldo Leopold. *A Sand County Almanac*. Oxford University Press, NY. 1966. 269 pp.

33. Amory Lovins. "Natural Economy." *Resurgence* 213 (2002):16-18.

34. For a more comprehensive discussion of how people think, see: Chris Maser. *Resolving Environmental Conflict: Toward Sustainable Community Development*. St. Lucie Press, Delray Beach, FL. 1996. 200 pp.

35. Karl F. Wenger. "Why Manage Forests?" *Journal of Forestry* 96 (January 1998): 1.

36. The preceding discussion of illegally cutting of trees on federal, public forests is based on: (1) The Associated Press. "Grant County residents vote on logging issue." *Corvallis Gazette-Times*, Corvallis, OR. May 20, 2002 and (2) John Enders. "Eastern Oregon pressure surfaces." *The Associated Press*. In: *Corvallis Gazette-Times*, Corvallis, OR. June 3, 2002.

37. Chris Maser. *The Perpetual Consequences of Fear and Violence: Rethinking the Future*. Maisonneuve Press, Washington, D.C. 2004. 373 pp.

38. John Perlin. *A Forest Journey: The Role of Wood in the Development of Civilization*.

W.W. Norton and Co., New York, NY. 1989. 445pp.

39. Theodore Roosevelt. *The Sunday Oregonian*, Portland, OR. July 22, 1990.

40. Gifford Pinchot. *The Training of a Forester*. J. B. Lippincott Co., Philadelphia, PA. 1914. 149 pp.

41. Society of American Foresters, "Forestry terminology." *Society of. American Foresters*. Washington, D.C. 1950. 93 pp.

42. International Union of Forestry Research Organizations. "Terminology of forest science, technology, practice, and products." Addendum Number One. *Society of American Foresters Publication*. 1977. 348 pp.

43. Edward Rothstein. "Paradise Lost: Can Mankind Live Without Its Utopias?" *The New York Times*. February 5, 2000.

44. I found the story of Flambeaux (which I have slightly modified) a long time ago, but can't remember where.

45. Victor Menotti. "free Trade, free Logging." *The International Forum on Globalization*. San Francisco, CA. Special Report. 1999. 30 pp.

46. (1) Satish Kumar and June Mitchell. "A Gandhian Future." *Resurgence* 211 (2002):44-48 and (2) John C. Gordon, Bernard T. Bormann, and Larry Jacobs. "The Concept of Ecosystem Fit and Its Potential Role in Forest Management: A Primary Research Challenge." *The Finnish IUFRO World Congress* (1996): 21-28.

47. Robert Rodale. "Big new ideas—where are they today?" Unpublished speech given at the Third National Science, Technology, Society (STS) Conference, February 5-7, 1988. Arlington, VA.

48. E. W. Sanderson, M. Jaiteh, M.A. Levy, and others. "The Human Footprint and the Last of the Wild." *BioScience* 52 (2002): 891-904.

49. The eulogy that Senator William Pitt Fessenden of Maine delivered on the death of Senator Foot of Vermont in 1866. *In*: John F. Kennedy. *Profiles in courage*. Harper & Row, New York, NY. 1961. 266 pp.

50. Pinchot, Gifford. *Breaking new ground*. Harcourt, Brace and Co., Inc., New York, NY. 1947. 522 pp.

51. Charles F. Wilkinson. *"The greatest good for the greatest number in the long run": The national forests in the next generation*. College of Forestry and Natural Resources, Colorado State University, Fort Collins, CO. Unpublished paper. 1985. 12 pp.

52. William McCall. "BPA chief returns $7,500 bonus." *The Associated Press*. In: *Corvallis Gazette-Times*, Corvallis, OR. July 3, 2002.

53. Chris Maser, Russ Beaton, and Kevin Smith. *Setting the Stage for Sustainability: A Citizen's Handbook*. Lewis Publishers, Boca Raton, FL. 1998. 275 pp

54. Edmund Burke. *In*: David W. Orr. "Conservatives Against Conservation." *Resurgence* 172 (1995):15-17.

55. Chris Maser. *The Perpetual Consequences of Fear and Violence: Rethinking the Future*. Maisonneuve Press, Washington, D.C. 2004. 388 pp.

56. The preceding discussion about the vision process that I conducted in Lakeview, OR, is from Dean Button, who used the results of my work as a case study for his doctorate dissertation: Dean Button. *Toward an Environmental Cosmology: The Power of Vision, Values, and Participation in Planning for Sustainable Development*. Ph.D. Dissertation. Antioch New England Graduate School. 2002. 154 pp.

57. For a discussion of a vision, goals, and objectives, see: Chris Maser. *Vision and Leadership in Sustainable Development*. Lewis Publishers, Boca Raton, FL. 1998. 235 pp.

58. Carl D. Holcombe. "1996 rainfall crushes old record." *Corvallis Gazette-Times*, Corvallis, OR. January 1, 1997.

59. For a thorough discussion of the biological functions in this example, see: pages 39-57 *In*: Chris Maser. *Sustainable Forestry: Philosophy, Science, and Economics*. St. Lucie Press, Delray Beach, FL. 1994.

60. S. Pyare and W. S. Longland. "Mechanisms of truffle detection by northern flying squirrels." *Canadian Journal of Zoology* 79 (2001): 1007-15.

61. For a discussion of ecological diversity and time scales, see: Chris Maser. *Ecological Diversity in Sustainable Development: The Vital and Forgotten Dimension*. Lewis Publishers, Boca Raton, FL. 1999. 401 pp.

62. (1) Wally W. Covington and M.M. Moore. *Changes in forest conditions and multiresource yields from ponderosa pine forests since European settlement*. Unplublished report, submitted to J. Keane, Water Resources Operations, Salt River Project, Phoenix, AZ. 1991. 50 pp.; (2) Gifford Pinchot. *Breaking new ground*. Harcourt, Brace and Co., Inc., New York, NY. 1947. 522 pp.; (3) Thomas W. Swetnam. "Forest fire primeval." *Natural Science* 3 (1988):236-241; (4) Thomas W. Swetnam. *Fire history and climate in the southwestern United States*. Pp. 6-17. *In*: Effects of Fire in Management of Southwestern Natural Resources. J. S. Krammers (Technical Coordinator). USDA Forest Service General Technical Report RM-191. Rocky Mountain Research Station, Fort Collins, CO. 1990.

63. (1) Larry D. Harris and Chris Maser. "Animal community characteristics." Pp. 44-8. In *The Fragmented Forest*. Larry D. Harris. Univ. Chicago Press, Chicago, IL. 1984. 211pp. and (2) Eric Sanford, Melissa S. Roth, Glenn C. Johns, and others. "Local Selection and Latitudinal Variation in a Marine Predator-Prey Interaction." *Science* 300 (2003):1135-1137.

64. The preceding discussion of insects is based on: (1) Timothy D. Schowalter. "Adaptations of insects to disturbance." pp. 235-386. *In*: S.T.A. Pickett and P.S. White (eds.). *The ecology of natural disturbance and patch dynamics*. Academic Press, New York. 1985; (2) Timothy D. Schowalter. "Forest pest management: A synopsis." *Northwest Environmental. Journal* 4 (1988): 313-318; (3) Timothy D. Schowalter. "Canopy arthropod community structure and herbivory in old-growth and regenerating forests in western Oregon." *Canadian Journal of Forest Research* 19 (1989): 318-322; (4) Timothy D. Schowalter, W.W. Hargrove, and D. A. Crossley, Jr. "Herbivory in forested ecosystems." *Annual Review of Entomology* 31 (1986):177-196; (5) Timothy D. Schowalter and Joseph E. Means. "Pest response to simplification of forest landscapes." *Northwest Environmental Journal* 4 (1988): 342-343; (6) Timothy D. Schowalter and Joseph E. Means. "Pests link site productivity to the landscape." pp. 248-250. *In*: David A. Perry, R. Meurisse, B. Thomas, R. Miller, and others (eds.). Maintaining the long-term productivity of Pacific Northwest forest ecosystems. Timber Press, Portland, OR. 1989; (7) David A. Perry. "Landscape pattern and forest pests." *Northwest Environmental Journal* 4 (1988): 213-228; (8) Timothy D. Schowalter. *Insect Ecology: an Ecosystem Approach*. Academic Press, San Diego, CA. 2000. 483 pp; (9) T.D. Schowalter and M.D. Lowman. *Forest Herbivory: Insects*. Pp 253-269. *In*: Ecosystems of Disturbed Ground (Lawrence R. Walker, editor). Elsevier, New York. 1999; (10) R.A. Progar and T.D. Schowalter. "Canopy arthropod assemblages along a precipitation and latitudinal gradient among Douglas fir *Pseudotsuga menziesii* forests in the Pacific Northwest of the United States." *Ecography* 25 (2002):129-138; (11) Timothy D. Schowalter with Jay Withgott. "Retinking Insects: What Would An Ecosystem Approach Look Like?" *Conservation Biology In Practice 2* (2001): 10-16; (12) T.D. Schowalter and L.M. Ganio. "Vertical and seasonal variation in canopy arthropod communities in an old-growth conifer forest in southwestern Washington, USA." *Bulletin of Entomological Research* 88 (1998): 633-640; (13) Robert A. Progar, Timothy D. Schowalter, and Timothy Work. "Arboreal Invertebrate Responses to Varying Levels and Patterns of Green-tree Retention in Northwestern Forests." *Northwest Science* 73 (1999): 77-86; and (14) Paul V.A. Fine, Italo Mesones, and Phyllis Coley. "Herbivores Promote Habitat Specialization by Trees in Ama-

zonian Forests." *Science* 305 (2004):663-665.

65. The preceding discussion of insect-eating birds is based on: John A. Weins. *Avian communities, energetics, and functions in coniferous forest habitats.* Pp. 226-265. In: Proceedings Symposium on Management of Forest and Range Habitats. D.R. Smith (editor). USDA Forest Service General Technical Report WO-1, U.S. Government Printing Office, Washington, D.C. 1975.

66. The preceding discussion of sugar maples is based in part on: F. Thomas Ledig and D. R. Korbobo. "Adaptation of sugar maple along altitudinal gradients: photosynthesis, respiration, and specific leaf weight." *American Journal of Botany* 70 (1983): 256-265.

67. The preceding discussion is based in part on: (1) Robert L. Park. "Scientists and Their Political Passions." *The New York Times*, New York. May 2, 1998. and (2) Mari N. Jensen. "CLIMATE CHANGE: Consensus on Ecological Impacts Remains Elusive." *Science* 299 (2003): 38.

68. John O. Stone, Gregory A. Balco, David E. Sugden, and others. "Holocene Deglaciation of Marie Byrd Land, West Antarctica." *Science* 299 (2003):99-102.

69. The preceding discussion of Alaska's melting glaciers is based on: (1) Mark F. Meier and Mark B. Dyurgerov. "Sea Level Changes: How Alaska Affects the World." *Science* 297 (2002): 350-351; (2) Michael D. Lemonick. "Life in the Greenhouse." *Time.* April 9 (2001): 24-29; and (3) Usha Lee Mcfarling. "Montana park's famous glaciers disappearing—quickly." *The Associated Press.* In: *Albany (OR) Democrat-Herald, Corvallis (OR) Gazette-Times.* November 24, 2002.

70. Françoise Gasse. "Kilimanjaro's Secrets Revealed." *Science* 298 (2002): 548-549.

71. (1) Per Gloersen, William J. Campbell, Donald J. Cavalieri, and others. "Satellite passive microwave observations and analysis of Arctic and Antarctic sea ice, 1978–1987." *Annals of Glaciology* 17 (2003):149–154 and (2) Josefino C. Comiso. "A rapidly declining perennial sea ice cover in the Arctic." *Geophysical Research Letters* (2002):(DOI: 10.1029/2002GL01650)

72. David J. Travis, Andrew M. Carleton, and R.G. Lauritsen. "Contrails reduce daily temperature range." *Nature* 418 (2002):601.

73. (1) Camille Parmesan and Gary Yohe. "A globally coherent fingerprint of climate change impacts across natural systems." *Nature* 421 (2003): 37–42 and (2) Terry L. Root, Jeff T. Price, Kimberly R. Hall, and others. "Fingerprints of global warming on wild animals and plants." *Nature* 421 (2003): 57–60.

74. The general discussion of mycorrhizae is based on: (1) David Read. "The ties that bind." *Nature* 388 (1997): 517-518; (2) David Read. "Plants on the web." *Nature* 396 (1998): 22-23; (3) Marcel G. A. van der Heijden, John N. Klironomos, Margot Ursic, and others. "Mycorrhizal fungal diversity determines plant biodiversity, ecosystem variability and productivity." *Nature* 396 (1998): 69-72; and (4) Anna S. Marsh, John A. Arnone, Bernard T. Bormann, and John C. Gordon. "The role of *Equisetum* in nutrient cycling in an Alaskan shrub wetland." *Journal of Ecology* 88 (2000): 999-1011.

75. This paragraph is based on: (1) Michael P. Amaranthus, Debbie Page-Dumroese, Al Harvey, and others. *Soil Compaction and Organic Matter Affect Conifer Seedling Nonmycorrhizal and Mycorrhizal Root Tip Abundance and Diversity.* Research Paper PNW-RP-494. U.S. Department of Agriculture, Forest Service, Pacific Northwest Research Station, Portland, OR. 1966. 12 pp; (2) Daniel L. Luoma. *Monitoring of Fungal Diversity at the Siskiyou Integrated Research Site, with Special Reference to the Survey and Manage Species* Arcangeliella camphorata *(Singer & Smith) Pegler & Young.* Unpublished Final Report, Order #43-0M00-0-9008. Department of Forest Science, Oregon State University, Corvallis, OR. 2001. 18 manuscript pages; and (3) J.E. Smith, R. Molina, M.M.P. Huso, and others. "Species richness,

abundance, and composition of hypogeous and epigeous ectomycorrhizal fungal sporo-carps in young, rotation-age, and old-growth stands of Douglas fir (*Pseudotsuga menziesii*) in the Cascade Range of Oregon, U.S.A." *Canadian Journal of Botany* 80 (2002):186-204.

76. Peter Högberg, Anders Nordgren, Nina Buchnamm, and others. "Large-scale forest girdling shows that current photosynthesis drives soil respiration." *Nature* 411 (2001): 789-792.

77. (1) Eric D. Forsman, E. Charles Meslow, and Howard M. Wight. "Distribution and Biology of the Spotted Owl in Oregon." *Wildlife Monographs* 87 (1984): 1-64; (2) Jack Ward Thomas, Eric D. Forsman, Joseph B. Lint, and others. *A Conservation Strategy for the Northern Spotted Owl: Report of the Interagency Scientific Committee to Address the Conservation of the Northern Spotted Owl.* U.S. Government Printing Office, Washington, D.C. 1990. 427 pp; (3) Andrew B. Carey, Janice A. Reid, and Scott P. Horton. "Spotted Owl Home Range and Habitat Use in Southern Oregon Coast Ranges." *Journal of Wildlife Management* 54 (1990): 11-17; and (4) Chris Maser. *Mammals of the Pacific Northwest: From the Coast to the High Cascade Mountains.* Oregon State University Press, Corvallis, OR. 1998. 406 pp.

78. (1) Zane Maser, Chris Maser, and James M. Trappe. "Food Habits of the Northern Flying Squirrel (*Glaucomys sabrinus*) in Oregon." *Canadian Journal of Zoology* 63 (1985): 1084-1088 and (2) Chris Maser, Zane Maser, Joseph W. Witt, and Gary Hunt. "The Northern Flying Squirrel: A Mycophagist in Southwestern Oregon." *Canadian Journal of Zoology* 64 (1986): 2086-2089.

79. (1) C.Y. Li, Chris Maser, and Harlan Fay. "Initial Survey of Acetylene Reduction and Selected Mircoorganisms in the Feces of 19 Species of Mammals." *Great Basin Naturalist* 46 (1986): 646-650 and (2) C.Y. Li, Chris Maser, Zane Maser, and Bruce Caldwell. "Role of Three Rodents in Forest Nitrogen Fixation in Western Oregon: Another Aspect of Mammal-Mycorrhizal Fungus-Tree Mutualism." *Great Basin Naturalist* 46 (1986): 411-414.

80. (1) Chris Maser, James M. Trappe, and Ronald A. Nussbaum. "Fungal-small mammal interrelationships with emphasis on Oregon coniferous forests." *Ecology* 59 (1978): 779-809, (2) Chris Maser, James M. Trappe, and Douglas Ure. "Implications of small mammal mycophagy to the management of western coniferous forests." *Transactions of the 43rd North American Wildlife and Natural Resources Conference* 43 (1978): 78-88, and (3) Daniel L. Luoma, James M. Trappe, Andrew W. Claridge, Katherine M. Jacobs, and Efren Cázares. *Relationships Among Fungi and Small Mammals in Forested Ecosystems.* Chapter 10. *In*: Cynthia J. Zable and Robert G. Anthony, (editors). Mammalian Community Dynamics in Coniferous Forests of Western North America: Management and Conservation. Cambridge University Press, New York, NY. 2002.

81. J. G. P. Calvo, Zane Maser, and Chris Maser. "A Note on Fungi in Small Mammals from the *Nothofagus* Forest in Argentina." *Great Basin Naturalist* 49 (1989): 618-620.

82. Andrew W. Claridge, M.T. Tranton, and R.B. Cunningham. "Hypogeal Fungi in the Diet of the Long-nosed Potoroo (*Potorous tridactylus*) in Mixed-species and Regrowth Eucalypt Stands in South-eastern Australia." *Wildlife Research* 20 (1993): 321-337.

83. The preceding discussion of decomposing wood is partly based on: (1) Chris Maser and James M. Trappe (Technical Editors.) *The Seen and Unseen World of the Fallen Tree.* USDA Forest Service General Technical Report PNW-164. U.S. Department of Agriculture, Forest Service, Pacific Northwest Forest and Range Experiment Station, Portland, OR. 1984. 56 pp., (2) Chris Maser, James M. Trappe and C.Y. Li. "Large Woody Debris and Long-Term Forest Productivity." *In: Proceedings of the Pacific Northwest Bioenergy System: Policies and Applications.* Bonneville Power Administration, May 10 and 11, Portland, OR. 1984. 6 pp., and (3) Chris Maser and James M. Trappe. "The Fallen Tree—A Source of Diversity." Pp. 335-339. *In*: Forests for a Changing World. *Proceedings of the Society of American Foresters*

1983 National Conference. 1984.

84. B.T. Bormann, H. Spaltenstein, M.H. McClellan, and others. "Rapid soil development after windthrow disturbance in pristine forests." *Journal of Ecology* 83 (1995):747-757.

85. (1) Michael H. McClellan, Bernard T. Bormann, and Kermit Cromack, Jr. "Cellulose decomposition in southeast Alaskan forests: effects of pit and mound microrelief and burial depth." *Canadian Journal of Forest Research* 20 (1990):1242-1246 and (2) Robert L. Deal, Chadwick Dearing Oliver, and Bernard T. Bormann. "Reconstruction of mixed hemlock-spruce stands in coastal southeast Alaska." *Canadian Journal of Forest Research* 21 (1991):643-654.

86. John J. Magnuson. "Long-term ecological research and the invisible present." *BioScience* 40 (1990): 495-501.

87. The following discussion of culverts is based on: "The Associated Press. Culvert fixes hamper salmon recovery." *Corvallis Gazette-Times*, Corvallis, OR, February 13, 1999.

88. E.M. Krümmel, R.W. Macdonald, L.E. Kimpe, and others. "Delivery of pollutants by spawning salmon." *Nature* 425 (2003): 255-256.

89. P. Bak and K. Chen. "Self-organizing criticality." *Scientific American*, January (1991): 46-53.

90. The preceding discussion of ecosystem fragility, and the example from ancient Greece, is based on: Fritz M. Heichelheim. *The effects of Classical antiquity on the land.* pp. 165-182. *In:* W. L. Thomas (Editor). Man's role in changing the face of the Earth. University of Chicago Press, Chicago, IL. 1956.

91. Eric Sanford, Melissa S. Roth, Glenn C. Johns, and others. "Local Selection and Latitudinal Variation in a Marine Predator-Prey Interaction." *Science* 300 (2003): 1135-1137

92. (1) F. Thomas Ledig, J. Jesús Vargas-Hernández, and Kurt H. Johnsen. "The Conservation of Forest Genetic Resources: Case Histories from Canada, Mexico, and the United States." *Journal of Forestry* 96 (1998):32-41 and (2) Lawrence H. Goulder and Robert N. Stavins. "Discounting: An eye on the future." *Nature* 419 (2002): 673-674.

93. The Holy Bible, Authorized King James Version. World Bible Publishers, Iowa Falls, IA. Numbers 35: 34.

94. The discussion about the importance of soil in this paragraph is based on: (1) G.C. Daily, P. A. Matson, and P. M. Vitousek. *Ecosystem services supplied by soil.* Pp 113-132. *In:* G. Daily, editor. Nature's Services: Societal Dependence on Natural Ecosystems. Island Press, Washington, D.C. 1997, (2) Gretchen C. Daily, Susan Alexander, Paul R. Ehrlich, Larry Goulder, Jane Lubchenco, and others. "Ecosystem Services: Benefits Supplied to Human Societies by Natural Ecosystems." *Issues in Ecology* 2 (1997): 1-16; and (3) David A. Perry. *Forest Ecosystems.* The Johns Hopkins University Press, Baltimore, MD. 1994. 649 pp.

95. L. R. Oldeman, V. van Engelen, and J. Pulles. 1990. The extent of human-induced soil degradation. Annex 5 of L. R. Oldeman, R. T. A. Hakkeling, and W. G. Sombroek, World Map of the Status of Human-Induced Soil Degradation: An Explanatory Note, rev. 2d ed. Wageningen: International Soil Reference and Information Centre.

96. The discussion of the formation of soil is based on: (1) Mark Ferns. "Geologic Evolution of the Blue Mountains Region, The Role of Geology in Soil Formation." *Natural Resource News* 5 (1995): 2-3,17; (2) Rob Marvin. "The Earth churns, moans, breathes, and the 'living rocks' keep rollin' on." *The Oregonian*, Portland, OR. April 18, 1991; (3) James L. Clayton. "Processes of Soil Formation." *Natural Resource News* 5 (1995): 4-6; (4) David D. Alt and Donald W. Hyndman. *Roadside geology of Oregon.* Mountain Press Publ. Co., Missoula, MT. 1978. 268 pp.; (5) Dwight R. Crandell. *The glacial history of western Washington and Oregon.* Pp. 341-353. *In:* The Quaternary of the United States. J.E. Wright, Jr., and David G. Frey (Eds.). Princeton Univ. Press, Princeton, NJ. 1965; (6) S.N. Dicken, and E.F. Dicken. *The*

making of Oregon: a study in historical geograph. Vol. 1. Oregon Historical Society, Portland. 1979. 207 pp.; (7) Alan E. Harvey. "Soil and the Forest Floor: What It Is, How It Works, and How To Treat It." *Natural Resource News* 5 (1995): 6-9; (8) Elaine R. Ingham. "Organisms in the Soil: The Functions of Bacteria, Fungi, Protozoa, Nematodes, and Arthropods." *Natural Resource News* 5 (1995): 10-12, 16-17; (9) Bernard T. Bormann, Deane Wang, F. Herbert Bormann, and others. "Rapid plant-induced weathering in an aggrading experimental ecosystem." *Biogeochemistry* 43 (1998): 129-155; (10) Michael Snyder. "Why is Soil Compaction a Problem in Forests?" *North Woodlands* 11 (2004): 19; and (11) A. G. Jongmans, N. van Breemen, U. Lundström, and others. "Rock-eating fungi." *Nature* 389 (1997): 682-683.

97. Matthias C. Rillig, Sara E. Wright, Michael F. Allen, Christopher B. Field. "Rise in carbon dioxide changes soil structure." *Nature* 400 (1999): 628.

98. Albert Tietema, Claus Beier, Pieter H.B. de Visser, and others. "Nitrate leaching in coniferous forest ecosystems: The European field-scale manipulation experiments NITREX (nitrogen saturation experiments) and EXMAN (experimental manipulation of forest ecosystems)." *Global Biogeochemical Cycles* 11 (1997):617-626.

99. Paul J. Squillace, Michael J. Morgan, Wayne W. Lapham, and others. "Volatile Organic Compounds in Untreated Ambient Groundwater of the United States, 1985-1995." *Environmental Science & Technology* 33 (1999): 4176-4187.

100. Elaine R. Ingham. "Organisms in the Soil: The Functions of Bacteria, Fungi, Protozoa, Nematodes, and Arthropods." *Natural Resource News* 5 (1995): 10-12, 16-17.

101. Mike Dorning. "Bush approves nuclear-waste site." *Corvallis Gazette-Times*, Corvallis, OR. February 16, 2002.

102. Leo Tumerman, quoted in *Corvallis Gazette-Times*, Corvallis, Oregon, 29 April, 1986.

103. Verne Gross Carter and Timothy Dale.*Topsoil and civilization, Revised Edition.* University of Oklahoma Press, Norman, OK. 1974.

104. W.C. Lowdermilk. "Conquest of the Land Through 7,000 Years." *Agricultural Information Bulletin No. 99*, U.S. Department of Agriculture, Soil Conservation Service, U.S. Government Printing Office, Washington, D.C. 1975. 30 pp.

105. Steve Newman. "Deforestation Alert." *The Los Angeles Times Syndicate.* In: *Albany (OR) Democrat-Herald, Corvallis (OR) Gazette-Times.* March 24, 2002.

106. The preceding discussion of soil is based on: (1) Andrew Sugden, Richard Stone, and Caroline Ash. "Ecology in the Underworld." *Science* 304 (2004): 1613; (2) Jocelyn Kaiser. "Wounding Earth's Fragile Skin." *Science* 304 (2004): 1616-1618; (3) Elizabeth Pennisi. "The Secret Life of Fungi." *Science* 304 (2004): 1620-1622; (4) J.R. McNeill and Verena Winiwarter. "Breaking the Sod: Humankind, History, and Soil." *Science* 304 (2004): 1627-1629; (5) David A. Wardle, Richard D. Bardgett, John N. Klironomos, and others. "Ecological Linkages Between Aboveground and Belowground Biota." *Science* 304 (2004): 1629-1633; (6) I.M. Young and J.W. Crawford. "Interactions and Self-Organization in the Soil-Microbe Complex." *Science* 304 (2004): 1634-1637; and (7) R Lal. "Soil Carbon Sequestration Impacts on Global Climate Change and Food Security." *Science* 304 (2004): 1623-1626.

107. The preceding discussion of carbon sequestration is based on: (1) David Adam. "Royal Society disputes value of carbon sinks." *Nature* 412 (2001): 108, (2) Erik Stokstad. "Defrosting the Carbon Freezer of the North." *Science* 304 (2004): 1618-1620, and (3) R Lal. "Soil Carbon Sequestration Impacts on Global Climate Change and Food Security." *Science* 304 (2004): 1623-1626.

108. The preceding three paragraphs are based on: (1) Andrew Balmford, and others. "Economic reasons for conserving wild nature." *Science* 297 (2002): 950-953; (2) Robert Costanza. "The value of the world's ecosystem services and natural capital." *Nature* 387

(1997):253-260; (3) Tomas Love, Eric Jones, and Leon Liegel. "Valuing the Temperate Rainforest: Wild Mushrooming on the Olympic Peninsula Biosphere Reserve." *Ambio Special Report* No. 9 (1998): 16-25; (4) Daniel Kahneman and Amos Tversky. "Prospect theory: an analysis of decision under risk." *Econometrica* 74 (1979): 263-291; (5) Peter H. Pearse. *Introduction to Forest Economics.* University of British Columbia Press, Vancouver, B.C., Canada. 1990. 226 pp; and (6) William A. Duerr. *Introduction to Forest Resource Economics.* McGraw-Hill, Inc., New York, NY. 1993. 485 pp.

109. Janet N. Abramovitz. "Learning To Value Nature's Free Services." *The Futurist* 31 (1997): 39-42.

110. Steve Newman. "Earthweek: A Diary of the Planet." *Albany (OR) Democrat-Herald, Corvallis (OR) Gazette-Times.* June 6, 1999.

111. Janet N. Abramovitz. "Learning To Value Nature's Free Services." *The Futurist* 31 (1997): 39-42.

112. Janet N. Abramovitz. "Learning To Value Nature's Free Services." *The Futurist* 31 (1997): 39-42.

113. David Ehrenfeld. "Obsolescence." *Resurgence* 193 (1999): 28-29.

114. The preceding discussion of landscape patterns is based in part on: Monica G. Turner. "Landscape ecology: The effect of pattern on process." *Annual Review of Ecological Systems* 20 (1989): 171-197.

115. Jerry F. Franklin and Richard T. T. Forman. "Creating landscape patterns by forest cutting: ecological consequences and principles." *Landscape Ecology* 1 (1987): 5-18.

116. Stephen J. Pyne. "Where Have All the Fires Gone?" *Fire Management Today* 60 (2000): 4-6.

117. David J. Rapport. "What constitutes ecosystem health?" *Perspectives in Biology and Medicine* 33 (1989): 120-132.

118. David J. Rapport, H.A. Regier, and T.C. Hutchinson. "Ecosystem behavior under stress." *American Naturalist* 125 (1985):617-640.

119. (1) Larry D. Harris *The fragmented forest.* University of Chicago Press, Chicago, IL. 1984. 211pp and (2) M. Rao, J. Terborgh, and P. Nunez. "Increased herbivory in forest isolates: Implications for plant community structure and composition." *Conservation Biology* 15 (2001):624-33.

120. (1) Peter H. Morrison and Frederick J. Swanson. *Fire History and Pattern in a Cascade Range Landscape.* USDA Forest Service General Technical Report. PNW-GTR-254. Pacific Northwest Research Station, Portland, OR. 1990. 77 pp. and (2) Charles Grier Johnson, Jr. *Vegetation Response after Wildfires in National Forests of Northeastern Oregon.* USDA Forest Service, Pacific Northwest Region. R6-NR-ECOL-TP-06-98. 1998. 128 pp.

121. (1) John Enders. "Oregon short on fire crews." *The Associated Press.* In: *Corvallis Gazette-Times,* Corvallis, OR. July 17, 2002 and (2) John Enders. "Weather adds to firefighting woes." *The Associated Press.* In: *Corvallis Gazette-Times,* Corvallis, OR. July 18, 2002.

122. Stephen F. Arno and Steven Allison-Bunnell. *Flames in Our Forest.* Island Press. 2002. 227 pp.

123. The preceding discussion of fire is based on: Andrew Kramer. "Oregon's two largest wildfires converge." *The Associated Press.* In: *Corvallis Gazette-Times,* Corvallis, OR. July 21, 2002.

124. James K. Agee. *Fire ecology of Pacific Northwest forests.* Island Press, Washington, D.C. 1993. 493 pp.

125. David A. Perry. "Landscape pattern and forest pests." *Northwest Environmental Journal* 4 (1988): 213-228.

126. Chris Maser, Robert F. Tarrant, James M. Trappe, and Jerry F. Franklin (Technical

Editors). *From the forest to the sea: A story of fallen trees*. USDA Forest Service General Technical Report PNW-229. Pacific Northwest Research Station, Portland, OR. 1988. 153 pp.

127. David H. Johnson and Tomas A. O'Neill (Managing Directors). *Wildlife-Habitat Relationships in Oregon and Washington*. Oregon State University Press, Corvallis, OR. 2001. 736 pp.

128. Chris Maser and James M. Trappe (Technical Editors.) *The Seen and Unseen World of the Fallen Tree*. USDA Forest Service General Technical Report PNW-164. U.S. Department of Agriculture, Forest Service, Pacific Northwest Forest and Range Experiment Station, Portland, OR. 1984. 56 pp.

129. (1) Chris Maser. Life histories and ecology of *Phenacomys albipes, Phenacomys longicaudus, Phenacomys silvicola*. M.S. Thesis. Oregon State University, Corvallis, OR. 1966. 221 pp; (2) Chris Maser. *Forest Primeval: The Natural History of an Ancient Forest*. Oregon State Univesity Press, Corvallis, OR , 1991. 282 pp.; (3) George C. Iverson, Gregory D. Hayward, Kimberly Titus, and others. *Conservation Assessment for the Northern Goshawk in Southeast Alaska*. USDA Forest Service General Technical Report. PNW-GTR-387. Pacific Northwest Research Station, Portland, OR. 1996. 101 pp; (4) Eric D. Forsman, Robert G. Anthony, and Cynthia Zable. *Distribution and Abundance of Red Tree Voles in Oregon Based on Occurrence in Pellets of Spotted Owls*. Unpublished manuscript on file at the USDA Forestry Sciences Laboratory, Corvallis, OR. 2003; (5) H.P. Hayes. *Arborimus longicaudus. Mammalian Species* 532 (1996): 1-5; and (6) N. Meiselman and Arlene T. Doyle. "Habitat and Microhabitat use by Red Tree Voles (*Phenacomys longicaudus*)." *American Midland Naturalist* 135 (1996): 33-42.

130. Jan Christian Smuts. *Holism and evolution*. MacMillan and Co., Ltd, London. 1926. 361 pp.

131. The preceding three paragraphs are based on: Ricarda Steinbrecher. "What is Wrong with Nature?" *Resurgence* 188 (1998): 16-19.

132. Antonio Regalado and Meeyoung Song. "Furor Over Cross-Species Cloning." *The Wall Street Journal*. March 19, 2002.

133. David Quist and Ignacio H. Chapela. "Transgenic DNA introgressed into traditional maize landraces in Oaxaca, Mexico." *Nature* 414 (2001): 541-543.

134. (1) David Pilz and Randy Molina, Editor. *Managing Forest Ecosystems to Conserve Wild Mushroom Harvests*. USDA Forest Service General Technical Report PNW-GTR-371. U.S. Department of Agriculture, Forest Service, Pacific Northwest Forest and Range Experiment Station, Portland, OR. 1996. 104 pp; (2) David Hosford, David Pilz, Randy Molina, and Michael Amaranthus. *Ecology and Management of the Commercially Harvested American Matsutake Mushroom*. USDA Forest Service General Technical Report PNW-GTR-412. U.S. Department of Agriculture, Forest Service, Pacific Northwest Forest and Range Experiment Station, Portland, OR. 1997. 68 pp; and (3) David Pilz and Randy Molina. "Commercial harvests of edible mushrooms from the forests of the Pacific Northwest United States: issues, management, and monitoring for sustainability." *Forest Ecology and Management* 155 (2002): 3-16.

135. The discussion of the cost of thinning national forests is based on: (1) Jeff Barnard. "Case study finds forest thinning expensive." *Corvallis Gazette-Times*, Corvallis, OR. August 22, 2002 and (2) Jeff Barnard. "Forest plan will face high costs, opponents." *Corvallis Gazette-Times*, Corvallis, OR. August 26, 2002.

136. (1) Matthew Daly. "Forest plan could log old-growth." *The Associated Press*. In: *Corvallis Gazette-Times*, Corvallis, OR. July 4, 2002 and (2) Matthew Daly. "Bush plans to ease logging rules." *Corvallis Gazette-Times*, Corvallis, OR. August 23, 2002.

137. (1) Jeffery Morton and Laura McCarthy. *A Comparison of Two Government Reports on Factors Affecting Timely Fuel Treatment Decisions*. Unpublished Report by the Forest Trust,

Santa Fe, NM. 2002. 8 pp. (Sent to Congress) and (2) Robert Gehrke. "Study: Enviro appeals no roadblock." *The Associated Press.* In: *Corvallis Gazette-Times,* Corvallis, OR. May 15, 2003.

138. Laurence McQuillan and Tom Kenworthy. "Bush defends forest plan as 'common sense' policy." *USA Today.* August 23, 2002.

139. (1) Jeffery Morton and Laura McCarthy. *A Comparison of Two Government Reports on Factors Affecting Timely Fuel Treatment Decisions.* Unpublished Report by the Forest Trust, Santa Fe, NM. 2002. 8 pp. (Sent to Congress) and (2) Robert Gehrke. "Study: Enviro appeals no roadblock." *The Associated Press.* In: *Corvallis Gazette-Times,* Corvallis, OR. May 15, 2003.

140. Tom Kenworthy. "Prevention efforts still missing mark after 2 years and $6 billion." *USA Today.* August 22, 2002.

141. Jerry F. Franklin, Kermit Cromack, Jr., William Denison, Chris Maser, and others. *Ecological Characteristics of Old-Growth Douglas fir Forests.* USDA Forest Service General Technical Report PNW-118. Pacific Northwest Forest and Range Experiment Station, Portland, OR. 1981. 48 pp.

142. David H. Johnson and Tomas A. O'Neill (Managing Directors). *Wildlife-Habitat Relationships in Oregon and Washington.* Oregon State University Press, Corvallis, OR. 2001. 736 pp.

143. (1) Andrew B. Carey, B.R. Lippke, and John Sessions. "Intentional systems management: managing forests for biodiversity." *Journal of Sustainable Forestry* 9 (1999): 83-125, (2) Andrew B. Carey, J. Kershner, and L.D. de Toledo. "Ecological scale and forest development: squirrels, dietary fungi, and vascular plants in managed and unmanaged forests." *Wildlife Monographs* 142 (1999): 1-71, (3) S. Duncan. Wisdom from the little folk: the forest tales of birds, squirrels, and fungi. *Science Findings* No. 16. USDA Forest Service, Pacific Northwest Research Station, Portland, OR. 1999, and (4) Jerry F. Franklin, Thomas A. Spies, R. Van Pelt, and others. "Disturbances and structural development of natural forest ecosystems with silvicultural implications, using Douglas fir forests as an example." *Forest Ecology and Management* 155 (2002): 399-423.

144. The preceding characterization of riparian areas is based on: (1) Jack Ward Thomas, Chris Maser, and Jon E. Rodiek. Riparian Zones. Chapter 3, pp. 40-47. *In:* Jack Ward Thomas (technical editor). Wildlife Habitats in Managed Forests: The Blue Mountains of Oregon and Washington. U.S.D.A. Forest Service Handbook No. 553. U.S. Government Printing Office, Washington, D.C. 1979 and (2) J. Boone Kauffman, Matthew Mahrt, Laura A Mahrt, and W. Daniel Edge. "Wildlife of Riparian Habitats." Pp. 361-388. *In:* David H. Johnson and Thomas A. O'Neil, Managing Directors. Wildlife-Habitat Relationships in Oregon and Washington. Oregon State University Press, Corvallis, OR. 2001

145. The preceding discussion of wood in riparian areas is based on: Chris Maser and James R. Sedell. *From the forest to the sea: The Ecology of Wood in Streams, Rivers, Estuaries, and Oceans.* St. Lucie Press, Delray Beach FL. 1994. 200 pp.

146. The preceding discussion of early England is based on: I.G. Simmons. "The earliest cultural landscapes of England." *Environmental Review* 12 (1988): 105-116.

147. The preceding discussion of the indigenous peoples of the Americas is based on: (1) Martin A. Baumhoff and Robert F. Heizer. *Postglacial climate and archaeology in the desert west.* pp. 697-707. *In:* The Quaternary of the United States. Wright, J.E., Jr., and D.G. Frey, (Eds.). Princeton University Press, Princeton, New Jersey. 1967; (2) James B. Griffin. *Late Quaternary prehistory in the northeastern woodlands.* pp. 655-667. *In:* The Quaternary of the United States. Wright, J. E., Jr., and D.G. Frey, (Eds.). Princeton University Press, Princeton, New Jersey. 1967; (3) Clement W. Meighan. *Pacific Coast archaeology.* pp. 709-720. *In:* The Quaternary of the United States. Wright, J. E., Jr., and D. G. Frey, (Eds.). Princeton University Press, Princeton, New Jersey. 1967; (4) Robert L. Stephenson. *Quaternary human occupa-*

tion of the plains. pp. 685-696. *In*: The Quaternary of the United States. Wright, J.E., Jr., and D.G. Frey, (Eds.). Princeton University Press, Princeton, New Jersey. 1967; (5) Stephen Williams and James B. Stoltman. *An outline of southeastern United States prehistory with particular emphasis on the Paleo-Indian Era*. pp. 669-683. *In*: The Quaternary of the United States. Wright, J.E., Jr., and D.G. Frey, (Eds.). Princeton University Press, Princeton, New Jersey. 1967; (6) Martyn J. Bowden. "The invention of American tradition." *Journal of Historical Geography* 18 (1992): 3-26, (7) Allan Chen. "Unraveling another Mayan mystery." *Discover*, June (1987):40,44,46,48-49; (8) K.A. Deagan. "La Navidad, 1492: searching for Columbus's lost colony." *National Geographic* 172 (1987): 672-75; (9) William M. Denevan. "The Pristine Myth: The Landscape of the Americas in 1492." *Annals of the Association of American Geographers* 82 (1992): 369-385; (10) W. George Lovell. "Heavy Shadows and Black Night: Disease and Depopulation in Colonial Spanish America." *Annals of the Association of American Geographers* 82 (1992): 426-443; (11) Philadelphia. "First Americans 30,000 years ago." *Corvallis Gazette-Times*, Corvallis, OR. February 17, 1998; (12) Richard L. Hill. "Digs in Peru and linguistics may prove humans lived in the Western Hemisphere 33,000 years ago." *The Oregonian*, Portland, Oregon. February 17, 1998; (13) Kim A. McDonald. "New Evidence Challenges Traditional Model of How the New World Was Settled." *The Chronicle of Higher Education*, March, 13 (1998):A22-A23; (14) S.M. Wilson. "That unmanned wild country: Native Americans both conserved and transformed New World environments." *Natural History*, May (1992): 16-17; (15) The Associated Press. "New Evidence suggests ancient immigrants came by boat." *Corvallis Gazette-Times*, Corvallis, OR. August 24, 1998; (16) Dr. Rob Bonnichsen, director for the Study of First Americans, Department of Anthropology, Oregon State University, Corvallis (personal communication); (17) Karl L. Butzer. "The Americas Before and After 1492: An Introduction to Current Geographical Research." *Annals of the Association of American Geographers* 82 (1992):345-368; (18) Douglas MacCleery. "Understanding the Role the Human Dimension has Played in Shaping America's Forest and Grassland Landscapes: Is There a Landscape Archaeologist in the House?" *Eco-Watch* 2 (1994): 1-12; and (19) Michael J. Heckenberger, Afukaka Kuikuro, Urissapá Tabata Kuikuro, and others. "Amazonia 1492: Pristine Forest or Cultural Parkland?" *Science* 301 (2003): 1710-1714.

148. (1) Stephen W. Barrett and Stephen F. Arno. "Indian fires as an ecological influence in the Northern Rockies." *Journal of Forestry* 80 (1982): 647-651; (2) James R. Habeck. "The original vegetation of the mid-Willamette Valley, Oregon." *Northwest Science* 35 (1961): 65-77; (3) Carl L. Johannessen, William A. Davenport, Artimus Millet, and Steven McWilliams. "The vegetation of the Willamette Valley." *Annals of the Association of American Geographers* 61 (1971): 286-302; (4) John T. Curtis. *The Vegetation of Wisconsin*. University of Wisconsin Press, Madison, WI.1959. 657 pp; and (5) Michael Williams. *Americans & Their Forests: A Historical Geography*. Cambridge University Press, New York, NY. 1989. 289 pp.

149. Robert V. Hine. *Community on the American Frontier: Separate But Not Alone*. University of Oklahoma Press, Norman. 1980. 292 pp.

150. The discussion of the bull trout is based on: (1) D.V. Buchanan, M.L. Hanson, and R.M. Hooton. *Status of Oregon' Bull Trout*. Oregon Department Fish and Wildlife Research and Reports 2003, (2) Washington Department of Fish and Wildlife. *Washington Bull Trout and Dolly Varden*. Fact Sheet. 1998, and (3) King County (Washington). *Ecology of the Bull Trout*. Fact Sheet. 2003.

151. The preceding discussion of the tropical rain forest is based on: Carol Savonen. *Ashes in the Amazon*. *Journal of Forestry* 88 (1990): 20-25.

152. Stephen J. Pyne. *Fire in America: A Cultural History of Wildland and Rural Fire*. University of Washington Press, Seattle, WA.1997. 654 pp.

153. Pinchot, Gifford. *Breaking new ground*. Harcourt, Brace and Co., Inc., New York,

NY. 1947. 522 pp.

154. George L. Hoxie. "How fire helps forestry." *Sunset* 34 (1910):145-151.

155. (1) Stephen J. Pyne. *Year of the fires: The story of the great fires of 1910.* Viking Press, NY. 2001. and (2) Michael Williams. *Americans & Their Forests: A Historical Geography.* Cambridge University Press, New York, NY. 1989. 289 pp.

156. Tom Kenworthy. "Prevention efforts still missing mark after 2 years and $6 billion." *USA Today.* August 22, 2002.

157. The preceding discussion about fire is based on: (1) Wally W. Covington and M.M. Moore. "Post-settlement changes in natural fire regimes and forest structure: Ecological restoration of old-growth ponderosa pine forests." *Journal of Sustainable Forestry* 2 (1994): 153-182; (2) Wally W. Covington and M.M. Moore. "Southwestern ponderosa pine forest structure: Changes since Euro-American settlement." *Journal of Forestry* 92 (1994):39-47; (3) Thomas W. Swetnam. "Forest fire primeval." *Natural Science* 3 (1988): 236-241; and (4) Thomas W. Swetnam. *Fire history and climate in the southwestern United States.* Pp. 6-17. In: Effects of Fire in Management of Southwestern Natural Resources. J. S. Krammers (Tech. Coord.). USDA Forest Service General Technical Report RM-191. Rocky Mt. Research Station, Fort Collins, CO. 1990.

158. Wally W. Covington and M.M. Moore. *Changes in forest conditions and multiresource yields from ponderosa pine forests since European settlement.* Unpublished report., submitted to J. Keane, Water Resources Operations, Salt River Project, Phoenix, AZ. 1991. 50 pp.

159. Carl G. Jung. *The Undiscovered Self.* A Mentor Book, New York, NY. 1958. 125 pp.

160. (1) B.P. Finney, I. Gregory-Eaves, M.S.V. Douglas, and J.P. Smol. "Fisheries productivity in the northeastern Pacific Ocean over the past 2,200 years." *Nature* 416 (2002): 729-733; (2) Keith R. Briffa and Timothy J. Osborn. "Blowing Hot and Cold." *Science* 295 (2002): 2227-2228; (3) Jan Esper, Edward R. Cook, and Fritz H. Schweingruber. "Low-Frequency Signals in Long Tree-Ring Chronologies for Reconstructing Past Temperature Variability." *Science.* 295 (2002): 2250-2253, and (4) Anthony R. Wood. "Experts: Climate may turn chilly." *Knight Ridder Newspapers.* In: *Corvallis Gazette-Times,* Corvallis, OR. December 9, 2002.

161. The Associated Press. "Oregon fires." In: *Corvallis Gazette-Times,* Corvallis, OR. July 30, 2002.

162. Jeff Barnard. "Agness residents ready to flee." *The Associated Press.* In: *Corvallis Gazette-Times,* Corvallis, OR. August 7, 2002.

163. The discussion of the Biscuit Fire is based on: (1) The Associated Press. "Fire largest on record." *Corvallis Gazette-Times,* Corvallis, OR. August 10, 2002; (2) The Associated Press. "Fire now largest on record." *Albany (OR) Democrat-Herald, Corvallis (OR) Gazette-Times.* August 11, 2002; (3) Jeff Barnard. "Spot fire force some temporary evacuations." *The Associated Press.* In: *Corvallis Gazette-Times,* Corvallis, OR. August 21, 2002; (4) The Associated Press. "Biscuit fire grows to become the largest in nation this year." *Corvallis Gazette-Times,* Corvallis, OR. August 24, 2002; (5) Jeff Barnard. "Biscuit Fire could be contained by Saturday." *Corvallis Gazette-Times,* Corvallis, OR. August 28, 2002; (6) Jeff Barnard. "Biscuit Fire gets no bigger for first time." *The Associated Press.* In: *Corvallis Gazette-Times,* Corvallis, OR. August 29, 2002; (7) Jeff Barnard. "Biscuit fire, largest in nation, contained." *The Associated Press.* In: *Albany (OR) Democrat-Herald, Corvallis (OR) Gazette-Times.* September 7, 2002; and (8) Jeff Barnard. "Nation's biggest and most expensive wildfire under control." *The Associated Press.* In: *Corvallis Gazette-Times,* Corvallis, OR. November 9, 2002.

164. (1) Tom Kenworthy. "Prevention efforts still missing mark after 2 years and $6 billion." *USA Today.* August 22, 2002, and (2) General Accounting Office (GAO). *Western national forests: A cohesive strategy is needed to address catastrophic wildfire threats.* U.S. General

Accounting Office, House of Representatives, Committee on Resources, Report to the Subcommittee on Forests and Forest Health RCED-99-65. Washington, D.C. 1999.

165. Robert Weller. "Too much at home in the forests." *The Associated Press.* In: *Albany (OR) Democrat-Herald, Corvallis (OR) Gazette-Times.* July 21, 2002.

166. The preceding historical perspective is based on: (1) L.S. Dillon. "Wisconsin climate and life zones in North America." *Science* 123 (1956):167-176; (2) R.L. Dix. *A history of biotic and climatic changes within the North American grassland.* pp. 71-89. *In:* D.J. Crisp (Ed.). Grazing in terrestrial and marine environments. Blackwells Science Publishing, Dorking, England. 1964; (3) E. Dorf. "Climatic changes of the past and present." *American Scientist* 48 (1960):341-346; (4) Ira S. Allison. "Fossil Lake, Oregon, its geology and fossil faunas." *Studies in Geology No. 9.* Oregon State University, Corvallis, OR. 1966. 48 pp.; (5) Martin A. Baumhoff and Robert F. Heizer. *Postglacial climate and archaeology in the desert west.* pp. 697-707. *In:* The Quaternary of the United States. H.E. Wright, Jr., and David G. Frey (Eds.). Princeton Univ. Press, Princeton, NJ. 1967; (6) James B. Griffin. *Late Quaternary prehistory in the northeastern woodlands.* pp. 655-667. *In:* The Quaternary of the United States. H.E. Wright, Jr. and David G. Frey (Eds.). Princeton University Press, Princeton, NJ. 1967; (7) Donald K. Grayson. " On the Holocene history of some northern Great Basin lagomorphs." *Journal of Mammalogy* 58 (1977):507-513; (8) Donald K. Grayson. "The biogeographic history of small mammals in the Great Basin: Observations on the last 20,000 years." *Journal of Mammalogy* 68 (1978):359-375; (9) Donald K. Grayson. *Mount Mazama, climatic change, and Fort Rock Basin archaeofaunas.* pp. 427-457. *In:* Volcanic activity and human ecology. Academic Press, Inc., New York, NY. 1979; (10) Guilday, J.E., P.W. Parmalee, and H.W. Hamilton. The Clark's Cave bone deposits and the late Pleistocene paleoecology of the Central Appalachian Mountains of Virginia. *Carnegie Museum of Natural History Bulletin* 2 (1977):1-87; (11) H.J.B. Birks and Hilary H. Birks. "The Rise and Fall of Forests." *Science* 305 (2004): 484-485; and (12) David A. Wardle, Lawrence R. Walker, and Richard D. Bardgett. "Ecosystem Properties and Forest Decline in Contrasting Long-Term Chronosequences." *Science* 305 (2004): 509-513.

167. Dennis Martinez. "Putting Ecology Into Restoration Forestry." *Distant Thunder, Journal of the Forest Stewards Guild.* 12 (2002): 1, 11.

168. Andrew N. Gray and Jerry F. Franklin. "Effects of multiple fires on the structure of southwestern Washington forests." *Northwest Science* 71 (1997): 174-185.

169. Chris Maser. *Forest Primeval: The Natural History of an Ancient Forest.* Oregon State Universityi Press, Corvallis, OR, 1989. 282 pp.

170. The preceding discussion of the biophysical effect of the Biscuit fire is based on: (1) Associated Press. "Biscuit fire didn't overcook." *Corvallis Gazette-Times,* Corvallis, OR. September 3, 2002 (2) Jeff Barnard. "Biscuit fire, largest in nation, contained." *Albany (OR) Democrat-Herald, Corvallis (OR) Gazette-Times.* September 7, 2002; and (3) Jeff Barnard. "What the Biscuit Fire left behind." *Albany (OR) Democrat-Herald, Corvallis (OR) Gazette-Times.* September 22, 2002.

171. Robert Gehrke. "Bush seeks to bypass courts in name of thinning." *The Associated Press.* In: *Corvallis Gazette-Times,* Corvallis, OR. September 6, 2002.

172. Dennis P. Lavender. "Burning is the wrong way to fire-proof the forests." Letter to the editor in the *Albany (OR) Democrat-Herald, Corvallis (OR) Gazette-Times.* September 8, 2002. (Note: In the following discussion, the sentences in italics are from this letter to the editor.)

173. (1) Jeffrey R. Waters, Kevin S. McKelvey, Daniel L. Luoma, and Cynthia J. Zabel. "Truffle production in old-growth and mature fir stands in northeastern California." *Forest Ecology and Management* 96 (1997):155-166 and (2) J.E. Smith, R. Molina, M.M.P. Huso, and

others. "Species richness, abundance, and composition of hypogeous and epigeous ectomycorrhizal fungal sporocarps in young, rotation-age, and old-growth stands of Douglas fir (*Pseudotsuga menziesii*) in the Cascade Range of Oregon, U.S.A." *Canadian Journal of Botany* 80 (2002): 186-204.

174. (1) B. T. Bormann and D.S. DeBell. "Nitrogen Content and Other Soil Properties Related to Age of Red Alder Stands." *Soil Science Society of America Journal*. 45 (1981): 428-432 and (2) Barnard T. Bormann. "A Masterful Scheme." *University of Washington Arboretum Bulletin*. 51 (1988): 10-14.

175. Bernard T. Bormann and John C. Gordon. "Can Intensively Managed Forest Ecosystems be Self-Sufficient in Nitrogen?" *Forest Ecology and Management* 29 (1989): 96-103.

176. (1) Jeff Barnard. "EPA study finds toxins in Columbia Basin Fish." *The Associated Press*. In: *Corvallis Gazette-Times*, Corvallis, OR. February 15, 2002, (2) Michelle Cole. "Pesticide reporting plan stalls." *The Oregonian*, Portland, OR. September 29, 2003, and (3) Fred Biddle and Jennifer Goldblatt. 2003. DuPont's troubled chemical [C-8]. *The News Journal*. November 23.

177. The discussion of salmon is based on: (1) Josef Hebert. "Farm runoff killing nation's coastal fish." *The Associated Press*. In: *Corvallis Gazette-Times*, Corvallis, OR, April 5, 2000, (2) The Associated Press. "Agricultural pollution worries area officials." *Corvallis Gazette-Times*, Corvallis, OR, November 9, 1999, (3) Scott Stouder. "Pesticides are overlooked part of salmon decline." *Albany (OR) Democrat-Herald, Corvallis (OR) Gazette-Times*. May 9, 1999, and (4) Jeff Barnard. "Environmentalists sue EPA over pesticides and salmon." *Corvallis Gazette-Times*, Corvallis, OR, January 31, 2001.

178. David Malakoff. "Arizona Ecologist Puts Stamp On Forest Restoration Debate." *Science* 297 (2002):2194-2196.

179. Stephen J. Pyne. "The Political Ecology of Fire." *International Forest Fire News* 19 (1998): 2-4

180. James K. Agee. "The landscape ecology of western forest fire regimes." *Northwest Science* 72(Special Issue) (1998): 24-34.

181. (1) General Accounting Office (GAO). 1999. Western national forests: A cohesive strategy is needed to address catastrophic wildfire threats. U.S. General Accounting Office, House of Representatives, Committee on Resources, Report to the Subcommittee on Forests and Forest Health RCED-99-65. Washington, D.C. and (2) Henry Carey and Martha Schumann. 2003. Modifying Wildfire Behavior—The Effectiveness of Fuel Treatments. National Community Forestry Center, *Southwest Region Working Paper 2*. 26 pp.

182. (1) Barnard T. Bormann, Martha H. Brookes, E. David Ford, and others. "Volume V: A framework for sustainable-ecosystem management." USDA Forest Service General Technical Report PNW-GTR-331. Pacific Northwest Research Station, Portland, OR. 1994. 61 pp. and (2) Barnard T. Bormann, Patrick G. Cunningham, Martha H. Brookes, and others. "Adaptive ecosystem management in the Pacific Northwest." USDA Forest Service General Technical Report PNW-GTR-341. Pacific Northwest Research Station, Portland, OR. 1994. 22 pp.

183. David Hulse, Stan Gregory, and Joan Baker (editors). *Willamette River Basin Planning Atlas: Trajectories of Environmental and Ecological Change*. Oregon State University Press, Corvallis, OR. 2002. 192 pp.

184. Bruce J. Peterson, Wilfred M. Wollheim, Patrick J. Mulholland, and others. "Control of Nitrogen Export from Watersheds by Headwater Streams." *Science* 292 (2001): 86-90.

185. (1) C. Jeff Cederholm, David H. Johnson, Robert Bilby, and others. "Pacific Salmon and Wildlife—Ecological Contexts, Relationships, and Implications for Management." Pp. 628-684. *In*: David H. Johnson and Thomas A. O'Neil, Managing Directors. Wildlife-Habi-

tat Relationships in Oregon and Washington. Oregon State University Press, Corvallis, OR. 2001 and (2) J. M. Helfield and R.J. Naiman. "Effects of salmon-derived nitrogen on riparian forest growth and implications for stream productivity." *Ecology* 82 (2001): 2403-2409.

186. (1) D.W. Schindler, K.G Beaty, E. J. Fee, D.R. Cruikshank, and others. "Effects of climatic warming on lakes of the central boreal forest." *Science* 250 (1990): 967-970; (2) Christopher Flavin. *Facing Up to the Risks of Climate Change.* pp 21-39. *In:* Lester R. Brown, Janet Abramovitz, Chris Bright, and others 1996. State of the World 1996: A Worldwatch Institute Report on Progress Toward a Sustainable Society. W.W. Norton & Co., New York, NY. 1966.

187. S. McCartney. "Watering the west, part 3. Growing demand, decreasing supply send costs soaring." *The Oregonian*, Portland, OR. 30 September, 1986.

188. L.C. Everard (editor). *Yearbook 1920.* United States Department of Agriculture, Government Printing Office, Washington, D.C. 1921. 888 pp.

189. The preceding discussion of water in California is based on: (1) Corvallis Gazette-Times. "California to lose river water rights." *Corvallis Gazette-Times*, Corvallis, OR. December 28, 2002 and (2) Seth Hettena. "Desert sea holds key to water future." *The Associated Press.* In: *Corvallis Gazette-Times*, Corvallis, OR. December 29, 2002.

190. Luna B. Leopold. "Ethos, equity, and the water resource." *Environment* 2 (1990): 16-42.

191. Steve Newman. Earthweek: "A Diary of the Planet (North China Subsidence)." *The Los Angeles Times Syndicate.* In: *Albany (OR) Democrat-Herald, Corvallis (OR) Gazette-Times.* September 15, 2002.

192. Mark W. Rosegrant, Ximing Cai, Sarah A. Cline. "Global water outlook to 2025: averting an impending crisis." (Food Policy Report) Washington, D.C., *International Food Policy Research Institute.* 2002. 36 pp.

193. D.J. Chasan. *Up for grabs, inquiries into who wants what.* Madrona Publishers, Inc., Seattle, WA. 1977. 133 pp.

194. (1) Alice Outwater. *Water: A Natural History.* Basic Books, New York, NY. 1996. 212 pp; (2) The Associated Press. "U.N. warning: Billions will face water shortages." *Corvallis Gazette-Times*, Corvallis, OR. March 23, 2002; and (3) Jim Carlton. "From Toilet to Tap: California Project Purifies Sewage Water." *The Wall Street Journal.* August 15, 2002.

195. T. Maddock, III, H. Banks, R. DeHan, R. Harris, and others. *Protecting the Nation's groundwater from contamination.* Washington, D.C.: U.S. Congress, Office of Technology Assessment, OTA-0-233. 1984. 244 pp.

196. W. J. Elliot, C.H. Luce, R.B. Foltz, and T.E. Koler. "Hydrologic and Sedimentation Effects of Open and Closed Roads." Blue Mountain Natural Resources Institute, La Grande, OR. *Natural Resource News* 6 (1996):7-8.

197. The preceding discussion of the stream-order continuum is based on: (1) Chris Maser and James R. Sedell. *From the forest to the sea: The Ecology of Wood in Streams, Rivers, Estuaries, and Oceans.* St. Lucie Press, Delray Beach FL. 1994. 200 pp, (2) Chris Maser. "The Humble Ditch." *Resurgence* 172 (1995): 38-40; and (3) David Stauth. "Streams may depend on violent floods, droughts." *Oregon State University News Service.* In: *Corvallis Gazette-Times*, Corvallis, OR. January 24, 2003.

198. Chris Maser. "Abnormal Coloration in *Microtus montanus.*" *Murrelet* 50 (1969): 39.

199. (1) Richard T. T. Forman and L. Alexander. "Roads and Their Major Ecological Effects." *Annual Review of Ecology and Systematics* 29 (1998): 207-231,(2) Stephen C. Trombulak and Christopher A. Frissell. "Review of Ecological Effects of Roads on Terrestrial and Aquatic communities." *Conservation Biology* 14 (2000): 18-30, and (3) Richard T.T. Forman. "Estimate of the Area Affected Ecologically by the Road System in the United States." *Conservation Biology* 14 (2000): 31-35.

200. (1) Jonathan L. Gelbrad and Jayne Belnap. "Roads as Conduits for Exotic Plant Invasions in a Semiarid Landscape." *Conservation Biology* 17 (2003): 420-432, (2) W. Schmidt. "Plant Dispersal by Motor Cars." *Vegtatio* 80 (1989): 147-152, and (3) David K. Person, Matthew Kirchhoff, Victor Van Ballenberghe, and others. *The Alexander Archipelago Wolf: A Conservation Assessment*. USDA Forest Service General Technical Report. PNW-GTR-384. Pacific Northwest Research Station, Portland, OR. 1996. 42 pp.

201. The Associated Press. "Logging-road problem bigger than thought." *In: Corvallis Gazette-Times*, Corvallis, OR. January 22, 1998.

202. The preceding discussion of the forest and the ocean is based on: (1) Chris Maser, Robert F. Tarrant, James M. Trappe, and Jerry F. Franklin (Technical Editors). *From the forest to the sea: A story of fallen trees*. USDA Forest Service General Technical Report PNW-229. Pacific Northwest Research Station, Portland, OR. 1988. 153 pp. and (2) Chris Maser and James R. Sedell. *From the forest to the sea: The Ecology of Wood in Streams, Rivers, Estuaries, and Oceans*. St. Lucie Press, Delray Beach FL. 1994. 200 pp.

203. B.G. Hatcher, R.E. Johannes, and A.I. Robertson. "Review of research relevant to the conservation of shallow tropical marine systems." *Oceanography and Marine Biology Annual Review* 27 (1989): 337-414.

204. Sir Charles Lyell. *Principles of geology; or the modern changes of the Earth and its Inhabitants*. D. Appleton & Co., New York, NY. 1866. 834 pp.

205. Chris Maser and James R. Sedell. *From the forest to the sea: The Ecology of Wood in Streams, Rivers, Estuaries, and Oceans*. St. Lucie Press, Delray Beach FL. 1994. 200 pp.

206. William E. Gibson. "So far, cost of U.S. war is a bargain." *South Florida Sun-Sentinel*. In: *Corvallis Gazette-Times*, Corvallis, OR. December 27, 2001.

207. Garrett Hardin. *Filters against folly: how to survive despite economists, ecologists, and the merely eloquent*. Penguin Books, New York, NY. 1986. 240 pp.

208. Lynn White, Jr. "The Historical Roots of Our Ecological Crisis." *Science* 155 (1967): 1203-1207.

209. Brother David Steindl-Rast. *Gratefulness And The Heart of Prayer: An Approach to Life in Fullness*. Paulist Press, Ransey, NJ. 1984. 224 pp.

###

New and Recent Books from Maisonneuve Press

Mas'ud Zavarzadeh and Donald Morton, eds., *Post-Ality: Marxism and Postmodernism*, 1995, 0-944624-27-8, $16.95.

Lenora Foerstel, ed., *Creating Surplus Populations: The Effects of Military and Corporate Policies on Indigenous Peoples*, 1996, 0-944624-31-6, $17.95.

Dennis, Crow, ed., *Geography and Identity: Exploring and Living Geopolitics of Identity*, 1996, 0-944624-23-5, $44.95 cl.; 0-944624-24-3, $19.95 pa.

Chezia Thompson-Cager, *The Presence of Things Unseen: Giant Talk*, 1996, 0-944624-32-4, $12.95.

Darko, Suvin, *Lessons of Japan: Assayings of Some Intercultural Stances*, 1997, 0-944624-37-5, $26.95.

Chris Maser, *The Perpetual Consequences of Fear and Violence: The, Rethinking the Future*, 2004, 0-944624-42-1, $19.95.

Chezia Thompson-Cager, ed., *When Divas Dance: The Diva Squad Poetry Collective*, 2004, 0-944624-43-X, $16.95.

Stephen Pelletiere, *Iraq and the International Oil System, Why We Went to War in the Gulf*, 2nd edition, 2004, 0-944624-45-6, $18.95.

Thomas Patrick Wilkinson, *Church Clothes: Or, Land, Mission, and the End of Apartheid in South Africa*, 2004, 0-944624-39-1, $46.95 cl.

Morse Peckham, *Man's Rage for Chaos, Biology: Behavior and the Arts*, 2005, 0-944624-36-7, $19.95.

Carl Mirra, ed., *Enduring Freedom or Enduring War: Costs and Prospects of the New American Century*, 2005, 0-944624-40-5, $19.95.

Herbert Foerstel, *From Watergate to Monicagate: Ten Controversies in Mass Media and Journalism*, 2nd. edition. 2005, 0-944624-44-8, $19.95.

Lenora Foerstel, ed., *The War on Children: Collateral Damage or Direct Policy in the War on Terrorism*, 2005, 0-944624-41-3, $19.95.

Victor Considerant, *Principles of Socialism: A Manifesto of the 19th Century*, 2005, 0-944624-47-2, $13.95.

Please request at your local bookstore or order directly from the press
Maisonneuve Press
P.O. Box 2980
Washington, DC 20013
http://www.maisonneuvepress.com
email to — orders@maisonneuvepress.com